GOAL PROGRAMMING TECHNIQUES FOR BANK ASSET LIABILITY MANAGEMENT

Kluwer Academic Publishers

Applied Optimization
Volume 90

Series Editors:

Panos M. Pardalos
University of Florida, U.S.A.

Donald W. Hearn
University of Florida, U.S.A.

GOAL PROGRAMMING TECHNIQUES FOR BANK ASSET LIABILITY MANAGEMENT

Kyriaki Kosmidou, Constantin Zopounidis

Technical University of Crete
Department of Production Engineering and Management
Financial Engineering Laboratory
University Campus, Chania, Greece

Kluwer Academic Publishers

Distributors for North, Central and South America:
Kluwer Academic Publishers
101 Philip Drive
Assinippi Park
Norwell, Massachusetts 02061 USA
Telephone (781) 871-6600
Fax (781) 871-6528
E-Mail <kluwer@wkap.com>

Distributors for all other countries:
Kluwer Academic Publishers Group
Post Office Box 322
3300 AH Dordrecht, THE NETHERLANDS
Telephone 31 78 6576 000
Fax 31 78 6576 474
E-Mail <orderdept@wkap.nl>

Electronic Services <http://www.wkap.nl>

Library of Congress Cataloging-in-Publication

Kosmidou, Kyriaki/ Zopounidis, Constantin
Goal Programming Techniques for Bank Asset Liability Management

ISBN Pb 978-1-4419-5475-6 e-ISBN E-book 978-1-4020-8105-7

Printed on acid-free paper.

Printed in the United States of America

To my sisters Marilena and Nadia Kosmidou

To Kalia Koukouraki, Dimitrios Zopounidis, Heleni Zopounidis

Benefactors love their beneficiaries more
than the beneficiaries love their benefactors

Aristotle (Ith., Eud., 1241a)

Table of contents

CHAPTER 2: REVIEW OF THE ASSET LIABILITY MANAGEMENT TECHNIQUES

CHAPTER 3: BANK ASSET LIABILITY MANAGEMENT METHODOLOGY

CHAPTER 4: APPLICATION

CHAPTER 5: CONCLUSIONS AND FUTURE PERSPECTIVES

Preface

Financial engineering involves the design, development and implementation of innovative financial instruments and processes and the formulation of creative solutions to problems in finance (Finnetry, 1988). Among others, financial engineering has been heavily involved in risk management, in assessing the types of risk of different securities, in identifying and measuring them, and finally in developing systems for transforming high risk investment means to low risk ones. Besides, risk management provides the most efficient way of managing risk through sophisticated quantitative and optimization models, such as *Asset Liability Management* (ALM) models.

In ALM the exposure to various risks is minimized by holding the appropriate combination of assets and liabilities in order to meet the firm's objectives. More precisely, allocating assets lies at the heart of a strategic risk management system. In addition, liability streams and their uncertainty, institutional constraints and policies, taxes, transaction costs and the like are important features in real financial planning. Application areas include pension plans, insurance companies, banks, university endowments and other leveraged institutions, wealthy and ordinary individuals. These investors possess future liabilities and goals. They must make investment decisions while considering the use of their funds, that is, investing for a purpose. Risks must be measured in the context of the entire organization's or the individual's financial situation. Asset investment decisions are combined with liability choices in order to maximize the investor's wealth over time.

During the last decades the growing internationalization, the globalization of financial markets, the intensified competition in the national and international banking markets and the introduction of complex products have increased volatility and risks. Within this economic environment, the banking institutions worldwide face new challenges to review their strategies, to proceed to technological innovations as well as to mergers and acquisitions.

The great and fast availability of all kinds of different information due to the development towards an "information society" has eliminated any delays between the occurrence of an event and the impact on the markets. Consideration of uncertainties is critical in financial planning. Moreover, the development of a stochastic model that takes into consideration the economic conditions, such as the deregulation of interest rate markets, is essential in the evaluation of the long-term investment strategies.

All the above have driven banks to seek out greater efficiency in the management of their assets and liabilities. Thus, the central problem of ALM revolves around the bank's balance sheet and the main question that arises is: what should the composition of a bank's assets and liabilities be on average given the corresponding returns and costs, in order to achieve certain goals, such as maximization of the bank's gross revenues? This need has led banks to determine the optimal balance among profitability, risk, liquidity and other uncertainties. The optimal balance among these factors cannot be found without considering important interactions that exist between the structure of a bank's liabilities, equity and the composition of its assets.

Taking into consideration all the above, the objective of this book is to provide a comprehensive discussion of the ALM problem as well as to review the existing methodologies. It also aims at the development of a bank ALM system into a stochastic environment, focusing, mainly, on the change of the interest rate risk. ALM is associated with the changes of the interest rate risk and specifically with the bond interest, deposit interest and loan interest, since the loans and deposits constitute the major accounts of the bank's balance sheet and the profitability sources of the banks. This ALM system provides the possibility to the administrative board and the managers of the bank to proceed to various scenarios related to their future economic process, aiming mainly at the management of the risks, emerging from the changes of the market parameters.

The book is organized in five chapters as follows:

Initially, in chapter 1 an introduction to ALM is presented. The general concepts related to ALM, its general model, as well as its applications are discussed. Moreover, a part of the first chapter is devoted to the financial institutions and the management of banking risks.

Chapter 2 provides a comprehensive review of existing ALM techniques. The review involves the discrimination of the methodological approaches to two basic categories, the deterministic and the stochastic techniques.

Chapter 3 is devoted to the development of the bank ALM system, that is based on goal programming methodology and simulation approach of the interest rate risk.

Chapter 4 presents the application of the developed system to a large commercial bank of Greece, analyses the results and proceeds to a forecasting analysis.

Finally, chapter 5 concludes the book, summarizes the main findings and proposes future research directions with respect to the study of the asset liability management problem in the banking field as well as in firms.

We would like to express sincere thanks to Dr Michael Doumpos, Lecturer at the Technical University of Crete for his valuable assistance in the preparation of the final manuscript.

Kyriaki Kosmidou Constantin Zopounidis

Chapter 1

Introduction

1. ASSET LIABILITY MANAGEMENT

Nowadays, because of the uncertainty and risk that exists due to the integrating financial market and technological innovations, investors often wonder how to invest their assets over time to achieve satisfactory returns subject to uncertainties, various constraints and liability commitments. Moreover, they speculate how to develop long term strategies to hedge the uncertainties and how to eventually combine investment decisions of asset and liability in order to maximize their wealth.

Asset liability management is the domain that provides answers to all these questions and problems.

More specifically, *Asset Liability Management* (ALM) is an important dimension of risk management, where the exposure to various risks is minimized while maintaining the appropriate combination of asset and liability, in order to satisfy the goals of the firm or the financial institution.

Up to the 1960's, liability management was aimless. In their majority, the banking institutions considered liabilities as exogenous factors contributing to the limitation of asset management. Indeed, for a long period the greater part of capital resources originated from savings deposits and deposits with agreed maturity.

Nevertheless, the financial system has radically changed. Competition among the banks for obtaining capital has become intense. Liability management is the main component of each bank strategy in order to ensure the cheapest possible financing. At the same time, the importance of decisions regarding the amount of capital adequacy is enforced. Indeed, the adequacy of the bank as far as equity, contributes to the elimination of bankruptcy risk, a situation in which the bank cannot satisfy its debts towards clients who make deposits or others who take out loans. Moreover, the capital adequacy of banks is influenced by the changes of stock prices in relation to the amount of the capital stock portfolio. Finally, the existence of a minimum amount of equity is an obligation of commercial banks to the Central Bank for supervisory reasons. It is worth mentioning that based on the last published data (31/12/2001) the Bank of Greece assigns the coefficient for the Tier 1 capital at 8%, while the corresponding European average is equal to 6%. This results in the configuration of the capital adequacy of the Greek banking system at higher levels than the European average rate. The high capital adequacy index denotes large margins of profitability amelioration, which reduces the risk of a systematic crisis.

Asset management in a contemporary bank cannot be distinct from liability management. The simultaneous management of assets and liabilities, in order to maximize the profits and minimize the risk, demands the analysis of a series of issues.

Firstly, there is the substantive issue of strategic planning and expansion. That is, the evaluation of the total size of deposits that the bank wishes to attract and the total number of loans that it wishes to provide.

Secondly, there is the issue of determination of the "best temporal structure" of the asset liability management, in order to maximize the profits and to ensure the robustness of the bank. Deposits cannot all be liquidated in the same way. From the point of view of assets, the loans and various placements to securities constitute commitments of the bank's funds with a different duration time. The coordination of the temporal structure of the asset liability management is of major importance in order to avoid the problems of temporary liquidity reduction, which might be very injurious.

Thirdly, there is the issue of risk management of assets and liabilities. The main focus is placed on the assets, where the evaluation of the quality of the loans portfolio (credit risk) and the securities portfolio (market risk) is more easily measurable.

Fourthly, there is the issue of configuration of an integrated invoice, which refers to the entire range of bank operations. It refers mainly to the determination of interest rates for the total of loans and deposits as well as for the various commissions which the bank charges for specific mediating operations. It is obvious that in a bank market which operates in a competitive environment, there is no issue of pricing. This is true even in the case where all interest rates and commissions are set by monetary authorities, as was the situation in Greece before the liberalization of the banking system.

In reality, bank markets have the basic characteristics of monopolistic competition. Thus, the issue of planning a system of discrete pricing and product diversification is of major importance. The problem of discrete pricing, as far as the assets are concerned, is connected to the issue of risk management. It is a common fact that the banks determine the borrowing interest rate on the basis of the interest rates which increase in proportion to the risk as they assess it in each case. The product diversification policy includes all the loan and deposit products and is based on thorough research which ensures the best possible knowledge of market conditions.

Lastly, the management of operating cost and technology constitutes an important issue. The collaboration of a well-selected and fully skilled personnel, as well as contemporary computerization systems and other technological applications, constitutes an important element in creating a low-cost bank. This results in the acquisition of a significant competitive advantage against other banks, which could finally be expressed through a more aggressive policy of attracting loans and deposits with low loan interest rates and high deposit interest rates. The result of this policy is the increase of the market stake. However, the ability of a bank to absorb the input of the best strategic technological innovations depends on the human resources management.

An analytical presentation of the general form of the asset liability management model, according to Ziemba and Mulvey (1998), is outlined below.

1.1 ALM model structure

Let us assume that the investment process consists of t={1,2,3,…,T} time periods, where t=1 is the current date and t=T is the planning horizon.

At the beginning of each period, the investor makes decisions regarding the asset mix, the liabilities and the financial goals. There are uncertainties between time periods. The primary decision variables designate asset proportions, liability related decisions and goal attainments, such as:

$x^s_{j,t}$ investment in asset *j* for time *t* and scenario *s*, where $s \in S$, S is a set of representative scenarios.

$y^s_{k,t}$ investment in liability *k* for time *t* and scenario $s \in S$

$u^s_{l,t}$ goal attainment *l*, for time *t* and scenario *s*.

In each time period *t*, the model maximizes its objective function, *f(x)*, by moving resources between asset categories, adjusting liabilities and achieving goals. In addition, constraints are imposed on the process such limiting borrowing to certain ratios, addressing transaction costs whenever assets are bought or sold or taking advantage of investment opportunities.

There are two basic equations for the flow of funds:

For the j asset, time t, and scenario $s \in S$, the following arises:

$$x^s_{j,t+1} = (x^s_{j,t} + r^s_{j,s}/100) - p^s_{j,t}(1+t_j) + q^s_{j,t}(1-t^+_j) \tag{1.1}$$

where

$r^s_{j,t}$ return for asset j for time t and scenario $s \in S$

$p^s_{j,t}$ sales of asset j for time t and scenario $s \in S$

$q^s_{j,t}$ purchase of asset j for time t and scenario $s \in S$

t_j transaction costs for asset j for time t and scenario $s \in S$

For the cash flows:

$$x^s_{l,t+1} = (x^s_{l,t} + r^s_{l,s}) - \sum_J q^s_{j,t} + \sum_j p^s_{j,t}(1-t^-_j) + w^s_t - \sum_k y^s_{k,t} - \sum_l u^s_{l,t} \tag{1.2}$$

where w^s_t are the cash flows at time t, scenario $s \in S$ and cash in asset category l.

The multi-stage investment model cannot optimize scenarios that do not represent a range of plausible outcomes for the future. To prevent this occurrence, non-anticipatory constraints are added to the model with the following form:

$x^{s_1}_{j,t} = x^{s_2}_{j,t}$ for all scenarios $s_1 \in S$ and $s_2 \in S$ that share a similar past up to time t.

The financial planning system addresses these non-anticipatory conditions, either explicitly or implicitly, and special purpose algorithms are available for solving the stochastic optimization model.

1.1.1 Objective functions

The general form of the objective function that calculates the profit (or loss) in the next year is:

$$\text{Expected profit} = \sum_{s \in S} f_s z^s \tag{1.3}$$

where f_s is the probability of scenario $s \in S$,

z^s is the profit or loss under scenario $s \in S$,

and S is a set of representative scenarios.

Generally, there are two primary theories for an objective function: the Von Neumann-Morgenstern theory and the classical utility function theory, as discussed bellow.

1.1.1.1 The Von Neumann-Morgenstern theory

The Von Neumann-Morgenstern expected utility maximizing theory (1944) remains the pre-eminent approach for making decisions under uncertainty. The optimization model is:

$$\max E(u(w)) = \sum_s f_s u(w^s) \tag{1.4}$$

where $u(w^s)$ is the Von Neumann-Morgenstern (VNM) preference function, for scenario $s \in S$, w^s is the investor's wealth under scenario $s \in S$, and f_s is the probability of scenario $s \in S$.

Once the solution w^* of VM is found, we determine its certainty equivalent value by computing the inverse function at the recommended solution, namely, $CE = u^{-1}(w^*)$, which represents the amount in cash that we would receive in order to sell (or buy) the random variable *w*.

Technically, it is convenient in many cases to assume that *u* is exponential and concave since for normally distributed scenarios the maximization of the preference function max *u(w)* is equivalent to the maximization of the mean-variance model: $\bar{\mu} - R_A \mu \sigma^2$

where R_A is the investor's Arrow-Pratt risk aversion index, namely $-u'' / u'$.

Although the VM theory provides a systematic basis for making consistent decisions, it does not address the temporal aspects of decisions over a planning horizon.

1.1.1.2 Classical utility functions

Most utility functions have little to do with risk premium and certainty equivalence as advocated by VNM. Instead, a utility function sets a numerical value dictating the relative importance for some characteristics of a model's performance. Since the investment model consists of discrete paths over the planning period with numerous variables of interest, the model's behavior should be summarized into a small group of statistical factors.

The range of plausible utility functions is large. For example, deviations from selected wealth targets could be penalized at selected time periods. Otherwise, a risk measure could be created, that would take into account asymmetries in the wealth function.

Penalties can be given substantial weights to reflect their importance: goals and liabilities which are time sensitive can be assigned higher priorities than goals which are less critical:

$$\max \varphi(x) = \lambda_1 g_1(x) + \lambda_2 g_2(x) + \ldots + \lambda_k g_k(x) \tag{1.5}$$

where $\varphi(x)$ is an utility function, $g_i(x), i = 1,2,\ldots,k$ is the i goal and $\lambda_i, i = 1,2,\ldots,k$ is the relative importance for goal i.

1.1.1.3 The Von Neumann-Morgenstern theory and utility functions

Several researchers have proved the precision of the VM theory with the intuitive appeal of classical utility functions.

Bell (1995) has shown that the utility function u: $u(w) = w - be^{-cw}$ for constants $b>0$, $c>0$ is equivalent to a modified risk measure. For a given investor, we could create efficient borders where risk would be based on outcome distributions that are asymmetric in shape. The investor's current position would be analyzed to make evaluations regarding the changes in risk and returns. All the above would yield an efficient utility risk-modified border.

The utility risk measurement is defined as follows:

$$Risk(wealth) = kx(\log(Expected\ value(e^{(wealth-w^*)})) \tag{1.6}$$

where $k = 2/(c^2)$, w^* is the expected wealth and c the coefficient in VM preference function.

This approach minimizes the effort of specifying the risk aversion coefficient. If the risk aversion index is specified then the functional form of the utility function does not differ significantly, at least in the case of normally distributed returns. Then, as shown in the calculations and theory of Kallberg & Ziemba (1983), the certainty equivalents and portfolio weights are similar in different utility functions if the average Arrow Pratt risk aversions are also similar.

1.2 Asset management models

Most researchers, in their effort to answer the question, «What should be the average composition of assets of a firm, given the corresponding return and cost, so that the firm can achieve certain goals such as the maximization of its income?», were led to various studies. The models that were developed regarding the optimal management of the assets of firms, the risk, the return and the liquidity are mainly:

- Stochastic Programming
- Decision Rules
- Capital Growth
- Stochastic Control.

In chapter 2, there is an extended review of the methodological approaches that were applied to the asset liability management, while in the present chapter the general forms of the models are presented briefly in order to point out their importance on a scientific and practical level.

1.2.1 Stochastic programming

Stochastic programming determines the optimal investment at each time period as a function of the expected company assets in comparison with the investor's circumstances using a large multi-stage stochastic program. A central idea is the generation of scenarios through a decision tree structure, as shown below:

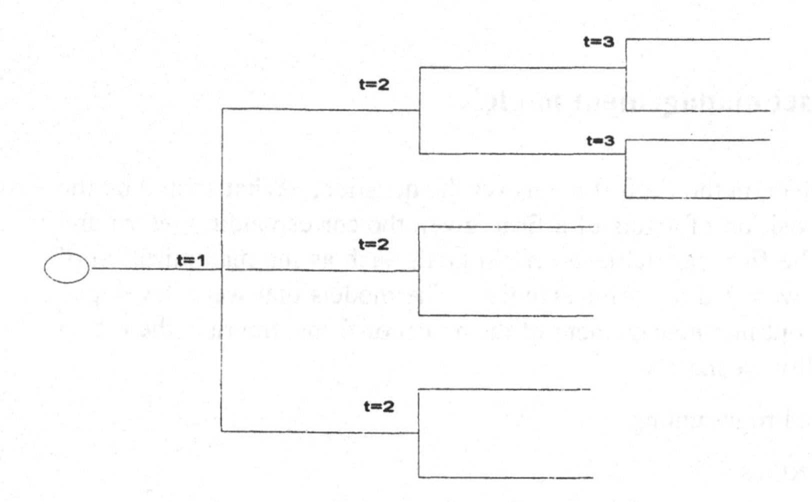

Figure 1.1: Structure of a decision tree

Stochastic programming models are based on the expansion of decision space, taking into account the conditional nature of the scenario tree. Conditional decisions are made at each node subject to modeling constraints.

An example of the stochastic programming approach was developed by Russell-Yasuda (Carino et al.,1994) at the Frank Russell Company. Their models use the following elements in various ways:

- Multiple time periods.
- Consistency with economic and financial theory.
- Discrete scenarios for random returns, liability costs and currency movements.
- Alternative forecasting models.
- Institutional, legal and policy constraints.
- Penalties for shortfalls in the objective function.
- Multiple expressions of risk.
- Maximized expected utility of final wealth net of penalty costs.

The Frank Russell Company has been active in developing such models. The general model ideas were developed by Kallberg, White & Ziemba (1982), and especially by Kusy & Ziemba (1986). More recent models were developed by Carino & Ziemba (1998) and Carino, Myers & Ziemba (1998).

An analytical presentation of these models follows in Chapter 2. A number of implementations also occurred in 1997, including an assets model for individual customers of the Banca Fidenram in Rome. Each of the models involved five periods including initialization and only the Russell-Yasuda model provided an efficient outcome.

The success of the models for institutional application demonstrates the practicality and importance of the approach. It is, thus, possible to successfully build and implement large-scale stochastic programming assets and asset-liability models.

1.2.2 Decision rules

A driving decision rule is a function for calculating the values of the investment and other business decisions at each time period. The data introduced is the state of the world at time t:

$$x_{j,t} = h\,(a^s_{j,t}, b^s_{j,t}, \ldots) \tag{1.7}$$

where $x_{j,t}$ is the investment in asset j for time t

and the parameters $a, b, \ldots$ depict the driving factors.

The decision variables are indexed over time and according to the state of the organization, but not necessarily over scenarios. A simple example is the fixed-mix strategy. At the end of each time-period, the investor sells over performing assets and purchases under performing assets to maintain a target level of the asset categories. The fixed-mix rule is as follows:

$$e_j = x^s_{j,t} / \Sigma x^s_{j,t} \tag{1.8}$$

where e_j is the fixed ratio for asset j

$x^s_{j,t}$ is the investment in asset j for time t and scenario s, where $s \in S$, S is a set of representative scenarios.

The amount of assets bought and sold can be computed by comparing the current portfolio and the ideal portfolio (as set by the e variables). The ratio of an asset as compared to total assets is fixed across all time periods and scenarios. The investor's surplus wealth may or may not be included in the evaluation. Even in this example, the investor must rebalance the portfolio at each stage. Mulvey and Chen (1996) show that this fixed-mix strategy reduces risks and improves returns over a passive buy and hold strategy.

It is relatively easy to develop decision rules to encompass an investor's wealth. For example, a common variant of portfolio insurance could be created called constant proportional portfolio insurance (CPPI-Constant Proportional Portfolio Insurance, Perold & Sharpe, 1988), which could modify the ratio of risky assets to risk-free assets as follows:

% Risky assets =min {D X (surplus wealth – F) + min Risk, max Risk} (1.9)

where D = risk aversion parameter,

F= minimum acceptable wealth level,

Surplus wealth= estimate of current position with respect to goals and liabilities,

minRisk = minimum level of risky asset and

maxRisk = maximum level of risky asset.

As the investor's wealth increases, the proportion of the risky assets increases with D. Aggressive investors will choose a large value (D=3), whereas conservative investors will choose a smaller value (D=0.3). Moreover, a limit is imposed on the percentage of the risky assets in the portfolio.

The time-step for implementing more aggressive CPPI strategies could be shorter than for the traditional fixed-mix strategy since it renders relatively large changes in the asset mix as the markets move. It is rather difficult to carry out this strategy during periods of high volatility and imbalances in supply and demand. The 1987 crash is an example where CPPI could have been proven inefficient since prices did not clear during substantial periods of time when the market experienced a sudden and sharp decline.

For CPPI, the definitions of the risky asset and the risk-free asset are general. For example, the risk free asset could be an immunized bond portfolio for some projected fixed liabilities. It could be purchased by inflation-linked government bonds and money market cash, when the liability cash flows grow with future inflation. Thus, as wealth increases, the investor will hold a higher percentage of riskier assets. The risk free asset could provide protection against adverse moves in the securities market.

Generally, the CPPI strategy performs best when markets are moving either up or down in a persistent way. The fixed-mix strategy performs best when markets are volatile but generally moving sideways.

Another approach for incorporating the investor's wealth into the decision rules is to define a target path over the planning horizon. The target wealth path should refer to the surplus wealth of the investor, as was proposed by Ziemba and Mulvey (1998):

Surplus wealth=total assets-PV (liabilities)-PV (goals) (1.10)

Where PV is the present value of future liability cash flows. The investor could be interested in maintaining a positive surplus wealth over time. In this dynamic environment, we develop an investment strategy and decision rules which control the relationship of the actual wealth to the target. For example, a more conservative choice, when the surplus wealth drives above the target path by a substantial level is also possible. It is possible to decrease or increase cash inflows. Alternatively, we can select the fixed-mix strategy that matches the target wealth as closely as possible, given the investor's tolerance for risks. The probability that the investor will reach a target wealth at a specified time could also be maximized.

Given one or more decision rules, we can build a multi-period ALM model for optimizing the setting of decision rules. For example, the best fix-mix proportions can be determined over the set of S scenarios. These optimization problems are relatively small but they often result in non-convex models and it is difficult to identify the global optimal solution. Examples of optimizing decision rules are Falcon Asset Liability Management (Mulvey, Correnti & Lummis, 1997) and Towers Perrin's Opt: Link System (Mulvey and Thorlacius, 1998).

1.2.3 Capital growth

Given a set of risky assets how should one invest to maximize the long run growth of assets? Kelly (1956) showed that under certain assumptions this is done by maximizing the expected log of asset wealth, that is, by using a logarithmic utility function. Breiman (1961) supplied rigorous mathematical proofs that this strategy did indeed asymptotically maximize long-run asset wealth and minimize the time to achieve a particular goal. Hakanson (1972) showed that the Kelly or capital growth strategy was myopic, that is to say optimization is optimal for general asset distributions according to each period. Rotand and Thorp (1992) applied the Kelly strategy to long-term investments in the US stock market and demonstrated some of the benefits and liabilities of that strategy. A major drawback is the size of the investment wagers on the most favorable assets leading to the highest growth rate.

1.2.4 Stochastic control

Dynamic stochastic control (SC) offers an alternative to stochastic programming for setting dynamic investment strategies. The approach dates back to the work of Samuelson (1969) and Merton (1969, 1990). A key idea is to form a state space for the driving variables at each time period. Rather than discretize the scenarios, stochastic control forms a mesh over the state space. Either dynamic programming algorithms or finite element algorithms are available for solving the problem. Brennan et al. (1997) and Brennan and Schwartz (1998) applied the stochastic control for asset allocation.

In order to incorporate liabilities in a dynamic stochastic control model, we must relate the present value of the liability cash flows to the driving economic variables. For example, a pension plan must pay beneficiaries over their retirement years and, in many cases, this increases payoffs as a function of inflation.

1.2.5 Advantages and disadvantages of the four approaches

Each of the four approaches to dynamic investment has something to offer. Decision rules are easier to implement and can be optimized without resorting to large scale linear or non-linear programs. They can be tested with out-of-sample scenarios and provide confidence limits on the recommendations. They are intuitive for most professional investors. However, they can lead to non-convex optimization models, requiring extensive research in finding a global optimal solution.

Stochastic programming provides a general purpose-modeling framework that can address real-world features such as turnover constraints, transaction costs, risk aversion, taxes, limits on groups of assets and other considerations. It requires highly efficient solution algorithms due to the large number of decision variables, especially for the multi-stage problems with 4 or more stages. Typical model applications from the Frank Russell research group include 5 stages. Its recommendations can be tested out-of-sample, but the computational costs are so high that they are impractical for many users. A limiting feature is the sampling of scenarios from the stochastic model.

Capital growth models lead to high growth of assets but with considerable risk. The models of asset generation, however, must be simple with liability considerations. They are relatively easy to solve and they tend to concentrate on a small number of superior assets and thus can be poorly diversified.

Stochastic control is another general-purpose framework and it applies to problems in which the state space can be kept manageable, with at least 3 or 4 driving variables. Other parameters and variables can be a function of the driving variables. Modeling errors may also arise due to the state space approximation. The difficulty in specifying general constraints on the process limits stochastic control applications. However, it has a conceptual edge over stochastic programming when it can be implemented since there is no need to sample scenarios.

In summary, there is no model that outperforms each of the others. It is suggested that investors start out with several candidate decision models and rules. They can be readily implemented and optimized. The selected decision rules can serve as benchmarks for the more complex stochastic programming approaches. Moreover, one can link multi-stage stochastic programming and decision rules to estimate confidence limits on the model's recommendations. Models that combine elements of the four approaches may also be desirable.

1.3 Applications of the asset liability management model

Applications of the asset liability management model could be found in insurance companies, banks, pension plans, portfolio and mutual fund managers, individuals, university endowments, etc.

- Pension plans

Actuaries evaluate the long-term viability of pension plans with respect to future contributions, anticipate pay-outs to beneficiaries and other future uncertainties. Legislation and changes in regulation, however, have led to the development of a methodology that measures the economic viability including risks in their studies.

- Insurance companies

Similar to pension plans, insurance companies are highly regulated and therefore their approach for analyzing a company's economic soundness is dictated by past regulations.

- Banks

Concerning the banking field, banks have been slow in implementing integrated risk management systems at a strategic level. The uncertainty of managing bank funds, based on profitability, liquidity, capital adequacy, investment return of legislative regulations (related to the stable against the changing interest rates as well as to risk factors) has led the banks to the efficient asset liability management.

- Portfolio and mutual fund managers

Many fund managers aim to beat a specified index such as the S&P 500, which is evaluated on its risk-adjusted performance compared to the benchmark index. An interesting example is Keynes' management of the trading assets (Chua & Woodward, 1983). In this context, the index equates to the liabilities and the investment goal is to compute risks relative to the return on the index. Portfolio managers place constraints on the investments. Asset categories can vary widely, whereas the number of decision variables increases as investment details are included.

- Individuals

Individuals can benefit by implementing dynamic asset and management strategies. They can evaluate the level of savings and investment strategies appropriate for meeting future financial goals, such as college education and retirement.

- University endowments

By nature, universities must consider a longer period of time when managing their endowment assets. Nevertheless, goals and future liabilities influence their investment risks. This is discussed in Merton's paper (1998). The basic idea is to locate investment opportunities that are closely correlated with liabilities and goals.

Before proceeding to the analytical presentation of the ALM methodologies, it would be appropriate to refer to a few characteristics of the banking institutions as well as to the uncertainty that prevails, since these are elements which determine the development of the methodology of the present research.

2. GENERAL CHARACTERISTICS OF THE BANKING INSTITUTIONS

Banking institutions constitute the core of the financing structure of a country. They accept deposits from various units and supply capital through borrowing and investments on individuals, firms and the government. Through these operations, they facilitate the flow of goods and services from the producers to the consumers, as well as the government funding activities. Thus, they contribute to the developing procedure of a country, while they constitute the means of the monetary policy's implementation. The banking system is therefore crucial for a country's economy.

Through the provision of funding services to the economy, their assets are almost of monetary form, while they include equity. They issue conventional liabilities to obtain capital which will ensure them more monetary funds. The net reserve funds of a credit institution, which emerge from the selling of securities or from the accumulation of earnings, represent a relatively small capital resource.

The efficient operation of banking institutions, in combination with the general economic goals, depends on their satisfactory management. As is the case for any institution or firm, banking institutions should have efficient management in order to avoid serious risks in the economy. They need to fulfill their goals if they are to develop a powerful, developing and evolving banking system, able to satisfy the demands of society.

2.1 The economic role of banking institutions

Financial institutions are firms that provide financial services to the economy through the financing form of their asset which is their basic characteristic and which distinguishes them from the other productive units.

Their role refers to the coverage of the needs of the borrowers and lenders in the economy. Their presence contributes to the accumulation of capital and the development of the economy in relation to the case of the economy where the money is used for transactions with the absence of financial institutions. This is accomplished due to their mediation in the savings-investment process which facilitates the differentiation between saving and investing decisions, the differentiation between ownership and management, while it encourages savings and investments resulting in larger growth rates than in the case where financial institutions do not exist.

The institution should set several goals, the most important of which refers to the structure of the asset and liability which will provide the desirable services to the various units.

The above, as well as other decisions over the management of the institution are based on the assumption that the main goal of the institution is the maximization of the stockholder's wealth. This maximization is consistent with the efficient distribution of resources, while the decisions of the institution take into consideration various regulations that have been instituted and refer to the social role of institutions.

According to the form that the asset and liability of the institution takes various alternatives are created. The institution is called to make choices among these alternatives based on the maximization of its stockholders' wealth. The practice of decision taking demands emphasis on several selected variables for the daily activities of the banking management. The most important differences between the management of the asset return and the liability charge are cost control, liquidity management and capital management.

The adequacy of the goal of the maximization wealth for the behavioral description of a few financial institutions has been argued, while other goals have been proposed, such as the maximization of the size of a banking institution.

The asset and liability of financial institutions are composed of various elements. There are various types of financial means and many methods of their classification. Moreover, there are particular methods of evaluation of the return for each of these elements, which, however, include various weaknesses, the most important of which refer to the precise risk of evaluation.

The decisions regarding the asset liability management should be based on the way they affect the risk and return of the asset portfolio of the banking institution rather than individual elements of this portfolio. Based on the assumption that the investor dislikes risk, it is possible to develop a model for the banking institution portfolio which will reflect the desirable combinations among risk and return.

2.2 Management of commercial banks

Commercial banks play an important role in the economy of a country since they affect the evolution of its monetary size as well as the developing procedure, specifically in developing countries, where they often constitute the unique accumulation mechanism of savings and funds and their distribution to the deficit units.

These abilities, from the banks' point of view, lead the authorities to the institution of specific regulations which aim to accomplish the desirable economic goals. These regulations involve demands of keeping liquid reserves at the banks in their capital condition, their liquidity, the deposit and loan interest rates, the allowance of loans for specific activities, as well as other types of transactions. These constraints affect banking decisions as well as the final configuration of the structure and size of their activities.

Commercial banks are organized in various forms, while the liability bank balance sheet is dominated by deposits on the liability side and loans on the asset side. The balance sheet structure constitutes one of the most important decisions of commercial banks, since it affects their income. However, banks should proceed to a series of other important decisions, which may become more difficult if uncertainty prevails in the financing markets.

The main goal of commercial banks, among all others, remains the maximization of the stockholders' wealth and having in mind the constraints that are imposed by the authorities, they formulate their portfolio in order to achieve it.

2.3 Basic policies of commercial banks

The basic policies of commercial banks relate to the collection of financial resources, borrowing, liquidity and capital.

The attraction of financial resources is significant for bank management, since each bank should have the required financial resources for the performance of its operations. Various administrative constraints and other restrictions affect the independence of banks in attracting capital. However, they are several factors that give the bank the possibility to affect its total volume. There are various means of finding capital, such as the various types of deposits, borrowing from the central bank and the purchase of various securities. Each bank should evaluate the capital cost of finding from various sources according to the flexibility they offer. In addition, it should take into account the factors which affect the capital in order to decide on their structure and ensure satisfactory return from the use of these funds. Furthermore, the competition which exists nowadays among various financial institutions in their aim to find financial resources should be underlined.

The borrowing of capital by any bank constitutes its primary operation and demands the execution of a well planned credit policy for the acquisition of a high return and the minimization of risk. The bank should decide for the size on the loans' portfolio, which should be determined by the needs of the economy and the capacity of the bank.

Moreover, it should decide on the distribution of its capital among the various sorts of loans, which differ in duration and risk and are affected by the environment, the borrowers' deposits at the given bank as well as other factors. A thorough market analysis must be made by each bank since there is no secondary market for most types of transactions. The experience from the specific customer, as well as the imperfections that prevail in the markets and generate uncertainties and risks should also be taken into consideration. Important decisions also concern the conditions that the bank imposes so as to lend the capital to the deficit units. Although the levels of interest rates are determined by the factors which the bank cannot affect, it has a degree of flexibility for the determination of its policy concerning its interest rates.

The latter are imposed on levels that cover payment for the use of money, the risk that is related to the borrower and other burdens related to the imperfections of the market. These factors are taken into account when dealing with various sorts of loans depending on the reliability of the customer. There are, nowadays, difficulties in determining loan interest rates although it is generally accepted that the least borrowing cost should constitute the lowest point in loan interest rates. Experience indicates that commercial banks follow the discount rate of the central bank, as well as the borrowing interest rate that they impose on their best customers, which they use as a basis for their other customers. Lately, the enforcement of fluctuant borrowing of interest rates is adopted, since the cost of a large proportion of bank loans is affected by the general level of interest rates.

Another basic policy of commercial banks refers to their liquidity management and specifically to the measurement of their needs related to the process of deposits and loans. For the coverage of this need, banks proceed to the management of liabilities, an activity which is, however, limited to large banks. The smallest banks are based on their assets – market of securities- which may imply loss of income for at least some time. Each bank should decide among the various sorts of securities concerning the quantity, the quality, the maturity and the return. The role of the central bank should also be emphasized since it acts as "the last banker" for commercial banks when covering their needs of liquidity.

The investment portfolio management for a commercial bank is important since the investments - securities etc - are vital as liquidity sources, income sources, while random investments lead to results which may not keep up with the desirable goals of the bank.

The risk factor of a bank plays an important role in the investment means and the qualitative levels of the bank portfolio. Thus, according to the capacity of a bank, the quality of securities obtained should be determined since they will influence its performance. The various securities have different characteristics in proportion to their quality - risk, return, uncertainty -. Moreover, the bank should make a decision concerning interest rates, the economic policy and the status of the economy. The selection of the above should be based on a degree of flexibility, in order to succeed in increasing income and profits.

Capital is crucial for a commercial bank especially for its immediate use for the extension of the bank credits and the attraction of financial resources, as well as for the provision of security - the protection of deposits - and the confidence needed for the promotion of bank operations. However, on the other hand, the large quantity of capital reduces the bank return. The rational policies try to balance the coverage provided by the extra capital to the public against the risk that exists in banking activities and the high return when more capital is involved.

2.3.1 The accumulation of capital

The accumulation of capital constitutes an important part of management policy in commercial banks. Most banks compete aggressively to attract more deposits, since the acquisition of capital is equally significant to capital management.

Commercial banks do not have complete control of deposits but do have the ability to affect the amount of deposits significantly. Many factors which determine the level of deposits such as monetary and fiscal regulations, as well as the general level of economic activity - get out of control. Banks could, however, influence an intermediate group of factors, such as the size and natural location of banks, while they could determine directly factors such as staff, promotion of their efforts and the level of services which they wish to offer to the depositors.

The desire of banks to borrow plays an important role in the accumulation of deposits. The availability of credit is substantial for most activities. When there is loss of capital, the banks prefer customers who maintain deposits in their branches. Credit facilitation is an important factor for the process of deposits. Banks are often ready to offer credit facilitation to firms which have no immediate need to borrow. At the same time these firms maintain their deposits with the banks, foreseeing the possible need to borrow capital in the future.

The savings deposits are another type of bank deposits. In this market, other non banking institutions can be active, such as the postal savings bank, where higher interest rates than those of commercial banks are allowed. In order to compete with these institutions commercial banks emphasize the wider flexibility which they offer with the application of various methods. The fact that commercial banks attract a significant amount of deposits, even when they compete with these institutions, proves the efficiency of their methods.

Another category refers to the deferred savings, which have a specific maturity date. Bank regulations allow the payment of higher interest rates in order to maintain the capital deposited for a longer period of time. This category is combined with the deposited amount. Thus, larger amounts obtain higher interest rates especially when they refer to long term time periods.

Transactions related to deposits can be differentiated more easily than those of asset funds, because they are often differentiated according to the funding means and the maturity date. The same funding means correspond to the demand characteristics related to them. However, uncertainty plays an important role. For example, the amounts repayable on demand are desirable for transaction reasons and thus those who ask for them are not concerned about the interest rates they pay either for the amounts repayable on demand or for other short-term types of accounts. Moreover, the saving deposits are used as a personal means of purchasing power or as a means of keeping funds for security reasons. If a rate of return is desirable for these funding means, then these means are indifferent to the interest rates, unless the interest rates that are paid differ significantly from the market interest rates for other short term accounts.

Uncertainty plays an important role in these transactions. This is probably more obvious in the transaction of the amounts repayable on demand. The capital from these accounts is forwarded upon demand and represents the most transitory part of liabilities in many banks. Historically, the movements to these accounts have been the most significant factor that leads several commercial banks to situations, where the assets should be converted under non-favorable prices, while few of them have caused significant difficulties.

Moreover, saving deposits also presuppose a degree of uncertainty which is related to the quantity of deposits. However, transition in these accounts is observed at certain time periods, in concordance with various regulations concerning interest rates that are assumed by monetary authorities. Such an example is the imposition of the highest limits in deposits or the administrative decisions for changes at these levels.

2.3.2 Loans

Loans constitute the most significant operation of commercial banks. A well planned loan policy is essential for the successful performance of the credit policy of a bank, for the acquisition of high return and the minimization of the risk that originates in credit extension.

A few characteristics of bank loans should be mentioned. Bank loans are significant in making deposits and attracting customers to other bank operations. The return from the loans is adjusted in order to compensate the bank for the risk related to the loan - credit risk, risk due to the loss of merchantability, risk of loans which present a stable interest rate - and also includes return from extra services provided to the borrowing customer.

Traditionally, banks follow a level of interest rates that derives from the discount rate imposed by the European Central bank. Although commercial banks classify the customers in proportion to the product and the branch, the market of the product and other factors, they have incomplete information related to the risk created by those who borrow as well as the position they hold in credit markets.

Loan conditions do not only include the interest rate that is imposed on the client but also the size of deposits which the bank demands from the borrower, during the program of quittance of the loan and the guaranties.

The effect of uncertainty, which is an important factor for banks, due to the default risk of loans, the changes in loan demand and the lack of sufficient knowledge over the flexibility of demand for the different categories of loans, should be emphasized once more.

2.3.3 Liquidity

Commercial banks should evaluate their needs of liquidity as well as the way these needs should be covered. These needs are determined by the difference between the increase of loan provisions and the development of deposits. These evaluations are based on the experience of the past and the various adjustments that should take place for various circular variations.

Banks should proceed with these evaluations since loans and deposits follow opposite directions in periods of high and low economic activity according to the government policy.

The preservation of a desirable level of liquidity is a painful procedure and demands specific dealings, which the banks often have difficulty to manage. The methodology that they usually apply refers to the liability management, that is, the acquisition of the liability in the market for the coverage of the demand for loans or the coverage of withdrawals and that of deposits. These activities refer to deposit acquisition, and often at high interest rates, from the capital market, deriving from the central bank. However, such a policy could be costly especially for small and intermediate size banks. These banks are based mainly on assets to satisfy the needs for liquidity and keep a sufficient amount of liquid funds. Although this method leads to income loss, banks avoid the possible high cost from the sales of junk bonds, especially in periods of significant expansion of economic activity.

2.4 Economic statements

Financial statements

Statements, which according to the law should be published, consist of the balance sheet accompanied with the income statement (profits-loss) and the disposal of net profits, as well as the statement of cash flows, which is obligatory in Europe, the United States and Canada and will also be obligatory in Greece with the implementation of IAS.

The balance sheet is a condensed statement of assets of the financial institution and the funding sources of these elements. The income statement is a condensed statement of income, expenses and net profit of the institution. In addition, the income statement always includes the analysis and the disposal of the net profit. It should be stated that notes or footnotes are added in the attachment at the end of the financial statements, that is the balance sheet and the income statement. Various methodologies and concepts are used as well as various significant facts that should be mentioned to the stockholders.

The balance sheet is a statement that presents the elements of Assets, Liability and Net Profit of a firm or a financial institution at a given date. Whatever transaction takes place after the editing of the balance sheet, the latter changes. The assets of a firm do not remain the same, but change in proportion to the size of the transaction that causes the change in the property of the firm. After the change, the new balance sheet is different from the previous one. The balance sheet presents the assets of a firm at a given time period and does not present the speculative ability of a firm, its profitability and efficiency. Thus, it constitutes a statement which presents the static image of the firm and not its dynamic structure. To comprehend the dynamic situation of the firm better, it is essential to use many consecutive balance sheets or the income statements in order to compare the profitability or non efficacy of the firm.

The balance sheet is a historical financial statement, because it presents the total transaction results of the past. Results of many profitable or not economic years are pointed out in the balance sheet. On the other hand the income statement presents the results of only one economic year and is not related to the past or the future.

Although the balance sheet is considered to be a financial statement of smaller importance than the income statement in making decisions of a business nature, it constitutes a basic statement and the information that it provides helps and facilitates the formulation of such decisions. This is especially necessary nowadays since inflation is a permanent phenomenon of our economy and the complex character of most decisions. For a more rational decision, we should use all the information that is available or could be available. Thus, the simultaneous use of all the financial statements and the accumulation of information from all the statements will lead to the best decision.

Assets or elements of assets have a value expressed in money which could be used to accomplish the objectives of the firm. These assets are recognized and evaluated according to the generally accepted concepts of accounting and IAS. A more dynamic definition supports that assets are those that provide the possibility of future economic benefits or services to the firm. The concept "Asset" means the future benefit or profit as well as the right of the firm to exploit these future perspectives.

The liability includes the short term and long term economic debts of a firm or a financial institution, which should be redeemed through the immolation of various assets or the provision of new liabilities. These are recognized and appraised according to the basic concepts of accounting and IAS. The liabilities must be generated through the transactions of the past and the present and need to be settled in the future.

The various accounts of the balance sheet of the banks according to the departmental accounting plan, the IAS and the accounting plan of the European Central Bank are mentioned to the following paragraph.

The Balance Sheet Statement

Asset

1. Cheques receivable and deposits at the Central Bank

This category includes the cash in euro and other currencies, deposits of the bank with the Central Bank, the balance of the current accounts in euro at the Central Bank and the balance of the accounts that the bank has at the Central Bank in foreign currencies, the value of the coupons that are for collection and finally the value of the cheques receivable through the office.

2. Treasury bills and other securities issued by the Central Bank for refunding

These include the interest-bearing treasury bills issued by the Greek state, separated in those that arise from the obligatory placement and those that are raised by the corresponding optional placements, as well as other securities that are at the bank's disposal. These securities, according to the current regulation of the monetary authorities, become accepted by the Central Bank for refunding.

3. Loans to credit institutions

These include repayable on demand loans as well as other loans and provisions for insecure demands.

4. Loans to customers

These include loans, other demands and provisions for insecure loans.

5. Securities

These include debt securities issued by the Greek state and by others as well as shares and other securities of variable return.

6. Participating interests

This category includes the participating interests in non-affiliated undertakings and the participating interests in affiliated undertakings.

7. Intangible assets

These include formation and preliminary expenses, other intangible assets and amortization of intangible assets.

8. Tangible assets

These include land, buildings, the equipment of buildings, furniture and fixtures, other tangible assets and payments on account and tangible assets.

9. Other assets

This category includes all the assets not incorporated into one of the above categories.

10. Prepayments and accrued income

These include deferred charges as well as accrued income of state bonds, other bonds and loans and advances.

Liability

1. Amounts owed to credit institutions

These include amounts repayable on demand and amounts with agreed maturity.

2. Amounts owed to customers

These include deposits repayable on demand, savings deposits and deposits with agreed maturity.

3. Other liabilities

4. Accruals and deferred income

These include deferred income, accrued interest on time deposits and other accrued expenses of the year.

5. Provisions for liabilities and charges

These include provisions for staff retirement benefits and other provisions.

6. Capital and reserves

These include stockholder's capital, that is the capital of the bank with reference to the number of stocks that it is divided, as well as the face value of each share and is distinguished to paid-up capital and to due capital. Moreover, this account includes the legal reserve, the extraordinary reserve, the tax-free reserves under special laws and the profit carried forward.

3. UNCERTAINTY IN THE BANKING RISK MANAGEMENT

As mentioned in the previous section, the uncertainty has an important role in the collection of capital and loans which constitute the basic part of the commercial bank asset liability management. The concept of financial risks is examined below as well as their evaluation and management techniques.

3.1 Risk of financial institutions

The financial risks are connected to the fluctuations of the financial markets. Naturally banks are involved more actively with financial risks, since they are affiliated directly with the object of their operation. The financial risks that a bank encounters are the market risk, the interest rate risk, the credit risk, the technological and operational risk, the foreign exchange risk, the country risk, the liquidity risk, the payment arrangement risk, the legal risk, the risk of reliability and the risk of insolvency. The efficient management of these risks is crucial to the return of the financial institutions.

The management of the financial risks has known great development during the last decades due to the instability that characterizes the international financial markets. There are many factors and events that have modulated the new unstable environment. The system of stable foreign exchange parities collapsed in 1971 leading to particularly changeable values of foreign exchange currency. The energy crisis that started in 1973, went on in 1979 and resulted in the consolidation of higher levels of inflation and the development of globalization in the decade of the 80's has led many banks of the Eastern Europe, which provided loans, to a difficult position. The development of the capital market of the developing countries in the 90's was accompanied by a severe crisis. However, in the mature markets the environment became more unstable. In October of 1987, the American capital markets sustained a 23% reduction with a capital loss of a value of one trillion dollars. In September of 1992, the process towards the monetary completion of Europe was at risk of being disconnected due to the breakdown of the European Monetary System. The values of the stocking titles in Japan were reduced significantly in the 90's with the Nikkei index slipping through from 39000 units to 17000 units three years later. The instability that characterizes the 90's was associated with a few bank collapses, such as the historical Barings Bank, in February of 1995, the Japanese Daiwa in September of 1995, BCCD in 1991, etc. The only international constant through all this period has been the high variability and the forecasting weakness.

As already mentioned in the previous sections, the traditional role of the financial institutions is the fixer between the deficit and the surplus economic units. That is, the financial system constitutes the operation core of the market economy converting the funds from savings to investment. In the contemporary financial environment, the role of the financial institutions has been more complex. The banks shift from the position of fixer to the position of financial risk manager.

The procedure of moving away from a fixer's position provides the firms with the possibility of their funding directly from the markets of capital and money. Thus, since the banks wish to maintain their profitability, they should be concerned with the assumption and management of risk. In this competitive environment, it is much easier to obtain profits.

The financial risk should be defined generally as the variability of the unexpected results of the markets of bonds, stocks and investment capital. The supervisory authorities of each country must comprehend the nature of these risks and investigate how the commercial banks measure and manage their efficiency. The general financial risk of a firm could be calculated as the variance or the standard deviation of the net income of a firm. At a bank that aims at the maximization of its profits, the calculation of risk should take place for the total of the bank or at the level of branches or services. The risk could be measured at the level of various banking products. In each case, the objective of the bank is to add value to its stock capital, maximizing the "adjusted to risk" returns of its stocks. In this sense, the bank behaves as any other firm. However, the profitability and the added value depend significantly on risk management. An insufficient risk management could threaten the solvency of the bank.

Market risk

The market risk derives from the uncertainty relating to the changes of the interest rates, financial values of foreign exchange parities and generally of the market parameters.

The market risk exists in the asset and liability collection due to the changes of the interest rates, the value of the foreign exchange and other values of assets. More specifically the market risk arises when the financial institution alternates assets and liabilities instead of keeping them for long term investment.

The market risk is distinguished into two categories (Jorion, 1997) depending on the form of relations that combine the various financial means. Firstly, the basic risk that is sustained when the relations among the financial products change. Secondly, the risk "gamma" that refers to the relations of non linear form among the products.

Another distinction of the market risk is based on the reasoning of the risk management strategy that follows. Thus, the market risk is distinguished to absolute risk, which is measured by the potential losses, such as dollars, euro and to relative risk, which is measured by a comparative index. Significant components of the market risk are the interest rate risk and the foreign exchange risk. An increase in the interest rates leads usually to the reduction of the bonds' values. A depreciation of the currency reduces the value of all the titles that are in this currency.

In financial theory, the market risk is defined as the variance of non anticipated results of the portfolio of titles due to the sudden variances of specific financial variables. In this sense, the positive as well as the negative deviations could be defined as risk sources. The public does not realize this fact and does not often recognize that the high returns of specific negotiator titles, such as Nick Leeson of Barings and Bob Citron of Orange County, include in reality high risks.

In order to measure the risk, the variable that interests the bank should be defined with accuracy. These variables could be the total value of the portfolio, the income, the capital or the returns of specific placements. The market risk refers to the influences of other financial factors, to the variables that is of interest. The risk is calculated by the standard deviation of this variable. The losses could arise by the combination of two factors: the variability of each financial factor and the degree of exposure to the changes of each factor.

The general or systematic market risk is affiliated with the variances of the values of all titles in the market due to an external factor or the expectations of the investors. The non systematic risk is affiliated with the value of a title, which operates in a different direction from the other titles of the market, due to the development that are related to the issuing institutions.

Interest Rate Risk

In mismatching the maturities of assets and liabilities as part of their asset transformation function, financial institutions potentially expose themselves to interest rate risk. An unexpected change in the interest rates could influence the profitability of a bank as well as the value of its stock. For example, assuming that in a bank its liabilities are more sensitive, in relation to its demands or to the changes of the interest rates, an increase in the interest rates could reduce the profits and a decrease of the interest rates could increase them.

In other words, the alteration of the asset elements includes the purchase of primary securities and the issue of secondary ones. The primary securities (bonds, exchanges, notes, e.tc.) that are usually bought from a financial institution are characterized by maturity and liquidity different from those of the secondary securities (stocks) that are available for sale. In matching the maturities of the assets and liabilities as part of the asset conversion function, the financial institutions are exposed to interest rate risk.

For the better comprehension of the interest rate risk the following example is presented: Consider a financial institution that issues liabilities of one-year maturity to finance the purchase of assets with a two-year maturity. Suppose the cost of funds (liabilities) for a financial institution is 9% per year and the interest return on an asset is 10% per year. Over the first year the financial institution can lock in a profit spread of 1% by borrowing short term (for one year) and lending long term (for two years). However, its profits for the second year are uncertain. If the level of interest rates does not change, the financial institution can refinance its liabilities at 9% and lock in a 1% profit for the second year as well. There is always a risk, however, that interest rates could change between the two years. If interest rates were to rise and the financial institution could borrow new one-year liabilities only at 11% in the second year, its profit spread in the second year would actually be negative (10%-11%=-1%). The positive spread earned in the first year by the financial institution from holding assets with a longer maturity than its liabilities would be offset by a negative spread in the second year. As a result, whenever a financial institution holds longer-term assets relative to liabilities, it potentially exposes itself to refinancing risk. This is the risk that the cost of rolling over or reborrowing funds could be more than the return earned on asset investments. On the contrary, if the financial institution holds shorter assets relative to liabilities, it is exposed to reinvestment risk.

In addition to a potential refinancing or reinvestment risk that occurs when interest rates change, a financial institution faces market value risk as well. It is known that the market value of an asset or liability is conceptually equal to the discounted future cash flows from the asset. Therefore, rising interest rates increase the discount rate on those cash flows and reduce the market value of that asset or liability. Conversely, falling interest rates increase the market values of assets and liabilities. Moreover, mismatching maturities by holding longer-term assets than liabilities means that when interest rates raise, the market value of the financial institution's assets falls by a greater amount than its liabilities. This exposes the financial institution to the risk of economic loss and insolvency.

Credit risk

Credit risk is the risk of default of accomplishment of a goal on the part of the borrower. The credit risk arises because promised cash flows on the primary securities held by financial institutions may or may not be paid in full. This insufficient response of the financial institution concerns the belated reimbursement of the institution to the performance of its obligations or the or the avoidance of its reimbursement. Each of these cases could be owed either to the inability of the financial institution to cover its liabilities or to the systematic avoidance of the confrontation of these liabilities by the financial institution.

Technology and Operational Risk

Operational risk refers to the losses that may emerge from the insufficient or unsuccessful internal procedure of the bank, bad operation of the systems, human errors, management failures as well as from other exogenous factors. Such problems may arise from the lack of preventive action.

An important sort of operational risk refers to the technology risk, that is the risks incurred by a financial institution when technological investments do not produce the cost savings anticipated. More specifically, technology risk occurs when technological investment does not produce the anticipated cost savings in economies of scale or scope. Diseconomies of scale arise because of excess capacity, redundant technology and organizational bureaucratic inefficiencies that get worse as a financial institution grows. Diseconomies of scope arise when a financial institution fails to generate perceived synergies or cost savings through major new technology investments. Other aspects of the operational risk include facts such as fires, earthquakes or other natural catastrophes.

Foreign Exchange Risk

The risk that exchange rate changes can affect the value of a financial institution's assets and liabilities located abroad is called foreign exchange risk. The foreign exchange risk is due to the fluctuations of the currencies' value that affect the "position" in exchange that a bank has undertaken for its funds' management or for its customers' satisfaction. The banks are acting in current as in deferred exchange purchase having large "positions" in exchange, that change continually.

Country risk

The country risk is defined as the probability of a country's failure to produce enough foreign exchange in order to serve its external debt (Cosset et al., 1992). More precisely, the country risk refers to the possibility of a country producing foreign exchange in order to serve its current and expected future debt. It includes the analysis of the present economic situation of a country and the forecasting of possible development giving more emphasis on the income of exportation, the expenses of importation and other data of the balance of current transactions of the country. Examining the present economic situation of a country and realizing that it shows a payment deficit, it is obvious that this country cannot satisfy its domestic demand for goods and services. Moreover, the concept of country risk is connected to the risk of the governmental interruption through prohibitions of payments abroad (sovereign risk), to the prohibition risk by the Central Bank to transfer foreign exchange abroad (transfer risk) and to the generalized risk.

Pay arrangement risk

The pay arrangement risk refers to the probability one of the two parties under contract to default the agreement, when the other party has already paid the money. This type of risk concerns the transactions dealing with foreign exchange, where the cash transfer from one bank to another is demanded, through the central banks whose currencies are used for the transaction. It becomes obvious that the arrangement risk is intense in the interbank market, where the volume and value of foreign exchanges are at high levels.

The evolution of the current payment systems contributes to the diminishment of the arrangement risk. Finally, the more efficient risk management demands the monitoring of the investment and other banking activities of the parties under contract.

Liquidity risk

The risk that a sudden surge in liability withdrawals may leave a financial institution in a position of having to liquidate assets in a very short period of time and at low prices is called liquidity risk. Liquidity risk arises whenever a financial institution's liability holders, such as depositors or insurance policyholders, demand immediate cash for their financial claims. When liability holders demand cash immediately, the financial institution must either borrow additional funds or sell assets to meet the demand for the withdrawal of funds.

Legal risk

The legal framework, which defines the bank operation, is possible to change, affecting the profitability of the banking institutions. A judicial decision that concerns a specific bank could have serious consequences on the arrangement of important banking issues. Moreover, the banks should investigate carefully the legal risk, when they develop new financial products or import new types of transactions.

The legal risk has often international dimension. The supervisory framework for banking activities differs among countries and could be interpreted differently. The wrong legal advice or defective legal documentation could lead to asset and liability deflation.

Credibility risk

The market risk and the credit risk, the country risk, the civil and foreign exchange risk constitute the basic core of the risks that banking institutions encounter and contribute to the change of the total value of their portfolio. The operational and legal risks belong to the group of operational risks that emerge from the nature of the bank as a business and economic unit. Finally, there is the risk of reputation and reliability, which is generated by the frequent failures of the operation systems, the bank management and the bank products. It is an important risk, since its presence undermines gradually the nature of the banking operations. The latter obviously demands the confidence of all the participants in the market.

Insolvency risk

The insolvency risk constitutes a consequence of the interest rate risk, of the market, foreign exchange, country, liquidity, credit risk as well as of the technological and operational risk. More specifically the insolvency risk is the risk that the financial institution does not have enough capital to encounter the eventual banking losses in the value of the assets in relation to the liabilities.

3.2 Evaluation and management risk techniques

The asset liability management depends on the interest rate risks in the market. It is claimed that the correct and successful management of financial risks and interest rate risk for banks and financial institutions contributes to their profitability. For this reason, however, the above institutions spend large amounts for their correct risk management.

The Capital Asset Pricing Model (CAPM), a basic theory in the risk management, presupposes the participation of all stocks/bonds in the portfolio. It recognizes the bisection of the risk to "non-systematic risk", where the risk disappears through the diversification of the investments and to "systematic risk", which arises by the controlled, exogenous factors, while mechanisms for its elimination do not exist.

Markowitz (1952), founder of the contemporary portfolio theory, supported that although the typical investor wants the maximum return, he also demands the minimum possible risk. According to Markowitz the investor should process the various alternative portfolios on the basis of the expected return and the variance using "indifference curves". The indifference curve represents the various combinations of risk and returns, which the investor finds equally desirable.

The efforts of the last decade have focused on the development of advanced means / systems that combine the various techniques, such as the value at risk analysis (VaR), the scenario analysis and the stress testing.

Value at Risk (VaR)

By the end of the 80's the application of risk quantification modes had begun. Their development is connected to the public debate that the market risk should be included in the Basel of Committee. In 1993, JP Morgan introduced the data bank "Risk Metrics" developing a method of market risk evaluation based on the approach of "Value at Risk" (Value at Risk – VaR, J.P. Morgan & Company, 1997). In 1996, the Basel of Committee encouraged the use of these models for the determination of the minimum supervisory capital demands against the market risk. More recently, many financial organizations, investment banks and mutual funds use the corresponding methods for the evaluations of their exposure to risks. Moreover, companies of evaluation of credit, such as Moody's and Standard and Poor's have announced their support of the VaR models.

The VaR model measures the quantity of capital of a financial organization that could be lost due to the variances of the portfolio value. That is, the issue is always the maximization of the current value of the portfolio through the control of the variance of cash inflows that are associated with the specific portfolio. The calculation of this quantity is based on a series of assumptions about the statistics of the time sequence of the value of each security as well as of the correlations among the values of the various securities.

Technically, when a financial institution keeps an "open place" in the market, for a given time period, VaR is the maximum possible loss that could occur with possibility y%. A typical definition of VaR could be expressed by the following equation:

$$\Pr(x \leq VaR) \succ y\% \quad \text{or} \quad \Pr(x \succ VaR) \leq (100 - y\%) \tag{1.11}$$

where Pr represents the possibility of the eventuality, x represents the real loss, VaR is the maximum value of the portfolio that could be lost in a specific time period under normal market conditions. Finally, an arbitrary period of confidence is represented by y. Assuming that for an "open place" y is 99%, there is at least 99% possibility for a real loss to be less than VaR. In other words, there is at least 1% possibility for the real losses to overcome the VaR quantity. The large advantage of VaR methodology is the fact that it incorporates in only one number the total exposure to the market risk of a financial institution. The simple and easy comprehension of this number explains why the VaR methodology has become so quickly an irreplaceable means for the presentation of the risk to the top executive staff, the administration and the stockholders of the attending risk. The application of VaR to programming allows the user to calculate the expected risk of a portfolio by the changes that arise in the values of foreign exchange / deferred payment, with a high level of validity from one to ten days. For example, the value VaR of a financial institution could indicate that the losses in the forthcoming week could surpass the 20 million euro with a possibility of not larger than 5%. If the administration estimates that the potential loss is particularly large, the financial institution must proceed to readjustment or hedging the total portfolio, in order to reduce the total Value at Risk (VaR).

The decision for the value of the three parameters is of the utmost importance for the planning of the model VaR. Firstly, the selected period of keeping depends on the frequency of the readjustments of the portfolio and the potential speed with which each financial institution could liquidate its places. Secondly, the width of the confidence interval fluctuates between the levels of statistical significance of 95% and 99%. The choice of the statistical significance level is indicative of each financial institution's behavior against the risk, that is one end of the spectrum is occupied by risk lovers and the other by risk aversion persons. Thus, the choice of a larger confidence interval (99% against 95%, significance level) reduces the possibility of the model VaR failing to forecast extreme events. Thirdly, there is the period that covers the sample of historical observations. For this period the variance and the covariance of the portfolio returns are calculated.

The advantage of historical samples of many observations is that it leads to a more precise evaluation of the real distribution of the returns. However, large samples are possibly available only for a limited number of risks in most industrialized and developed countries. More specifically, the choice of a width of a time sequence could satisfy two contradictory demands. On one hand, the largest the number of observations is the more precise the evaluation of the risk. On the other hand, the behavior of the time sequence changes in the long run due to its stochastic nature. Since there are temporal turning points, after which the value of the structural parameters of the model changes, the use of the information, incorporated in the calculation of the variability before the turning point for the evaluation of the future variability reduces rather than reinforces the forecasting ability of the model. Thus, the historical sample that maximizes the forecasting ability of the model is the one that includes the period after the last turning point.

Changes to the three parameters of the model could lead to various evaluations of the risk. This problem is known as "model risk", meaning the risk of evaluating the positions that the financial institution "opens" via a model that suffers either from wrong specialization or by inefficient estimation of the parameters. For the confrontation of the "model risk" the grading of the model through real data is necessary. The Basel of Committee defines a green, a yellow and a red zone, depending on the number of cases where the losses overcome the daily estimates of VaR. The green zone concerns the models whose grading detects one to four violations. The yellow zone concerns those with five to nine violations. Finally, the red zone concerns the models with more than ten violations. In the last case, the supervisory authorities could demand from the bank either to ameliorate its model or to withdraw it. However, the classification techniques of the models, based on the grading, have been sustained to serious criticism (Kupiec, 1995). More specifically, these techniques do not have statistical vigor to discriminate among the "good" and "bad" models, when the time horizons of the grading are narrow, such as the minimum of a year that the Basel of Committee forecasts.

The possibilities of VaR approach have, however, their limits. A simple VaR model is based on the assumption that the changes in the value of the portfolio data are normally, independently and diachronically distributed, whereas the correlations and the variances change slowly. The VaR models tend to assume that the structure of the market is simple and stable. The VaR value depends on the (more in the case of the historical simulation method and the variance-covariance analysis and less in the case of Monte Carlo method) hypothesis that the future will adjust to the present data. As it is obvious from the recent financial crisis, these models tend to under evaluate the possibility of extreme data.

At the same time, it is acceptable that the behavior of the returns in the financial markets is described only partly by the normal distribution. In practice, the empirically observed frequency distributions in the financial markets have large outcomes. In other words, large variances in the market take place more often than the normal distribution forecasts (Jackson et al., 1997).

Stress testing

In contrast to the VaR approach, stress testing and the scenario analysis do not give emphasis to the normality of statistical observations, but to the skewness of their frequency distribution.

The stress testing method aims at the inspection of the effects, which the change of specific factors could have on the portfolio value of a financial institution.

The stress testing method deals with the effects of such outliers. It only examines the large changes of specific variables, which barely deal with the daily observation of the risks, but could happen. The risk management unit in a financial institution should follow the stages of the stress testing process below.

First, the identity stage. That is it should define the disturbances that the analysis will include and their size. It is possible to examine, for example, disturbances such as a parallel or not transposition of the return curve, a reduction of the liquidity, a bankruptcy of the specific title or a group of titles of a geographical region or a branch.

Second, the definition of frequency through which the analysis of the stress testing will take place.

Third, the formulation of a hypothesis series that could affect the structure of the analysis and the interpretation of the results. Indeed, if there are serious empirical findings that support the existence of a powerful correlation between the two significant variables, the hypothesis that the initial disturbance will expand to a "family of variables" is likely to take place.

On the other hand, one could support that it is preferable not to assume the above, because a powerful disturbance is possible to lead to a collapse of correlations and parameters of the structural model that connects the variables. It is essential to clear the hypothesis whether the existed correlations collapse or stand. For the correlations that subside, new hypotheses are demanded. Finally, it is possible to assume that such case history could happen in other countries, industrial branches, etc or to consider that the correlations are non-existent.

Scenario analysis

The scenario analysis and stress testing examine a hypothetical change of the present situation calculating its effects on the bank. There are, however, significant differences between the scenario analysis and stress testing. Stress testing provides information about the situation of the title portfolio, if one or more variables go towards a specific direction and at a specific degree. On the other hand, the scenario analysis starts with a detailed analytical framework regarding the future dominance of alternative situations and explores the process of the title portfolio and the rest of the financial data in each case. It uses multidimensional forecasts and helps the financial institution to determine its long term strategic inabilities.

On the contrary, stress testing is a one-dimensional forecasting method, since it considers the effect of change of the economic environment on the bank. If both methods provide the bank with the possibility to determine and confront problems that may arise in the future, stress testing examines the short term effects of the change of a factor, whereas the scenario analysis examines in long term the consequences of more complicated developments in the economic environment. It becomes obvious that stress testing aims mostly at the solution of policy problems, whereas the scenario analysis at the confrontation of strategy matters of a financial institution.

The stages of the scenario analysis are four. The first stage concerns the determination of a "reasonable scenario". Moreover, it is often valuable to examine other scenarios, so that the analyst may utilize their comparison. The two principles that should be taken into account during the choice of appropriate scenarios are the complete knowledge of all the characteristics of the portfolio and the comprehension of the developments in the financial markets.

The second stage, known as "field analysis", aims at the determination of all the financial risks that are affected by the "reasonable scenario". This stage, besides the quantitative analysis, demands a procedure of interviews with all the involved management directions and departments of the bank, in order to refine the scenario and to access the statistical data.

The third stage constitutes the core of the scenario analysis, since forecasting in each "field" separately takes place and the forecasting ability is examined.

Finally, during the fourth stage the individual forecasts for each field are integrated in a fully complete scenario. The scenario is examined regarding problems of internal consistency, double calculation or contradictory assumptions.

Concluding, according to Mulvey et al. (1997), the choice of scenarios for asset liability management should take place so that it may represent a series of results concerning the values of the objective function. Moreover, those events that contribute to the optimal value of the expected utility function should be included with the possible scenarios. If the decision maker expresses aversion towards the risk, a number of scenarios that lead to the minimum return should be selected. The model, tries to avoid or at least to minimize the effect of scenarios through the strategies of asset data distribution.

Interest rate risk management techniques

The interest rate change is a strategy of the central bank. If the Central Bank smoothes the levels of interest rates, the unexpected disturbances and the variability of the interest rates tend to be less. Moreover, the exposition to risk of a financial institution due to the fact that the maturity dates of the asset and liability data may not coincide tends to be small.

The variability of the interest rates and the risk that the Central Bank could attribute to a status of swiftly converted reserve-targeting regime places the measurement and the interest rate risk management to the top of problems that the current managers of the financial institutions encounter.

The banks, in general, use various methods, which determine the effect of net profits before tax by the interest rate changes. For example, if a firm maintains its asset and liability invariable, a change in the interest rates affects the net profit before tax only through a change in the payments of interest rates of the existent asset and liability. As the interest rates change, only few asset and liability data are influenced during a period.

Conventional conditions determine which assets and liabilities replicate more than a period. For example, the interest rate payments to outstanding mortgages with stable interest rates do not replicate, but the interest rate payments to specific deposits with short-term maturity dates replicate quickly. The net outcome is that the bank with large mortgage portfolio of stable interest rate, which is financed by the deposits of short-term maturity encounters a fall in the interest rate income since the interest rates increase.

A bank whose liability replicates faster than the assets is called liability sensitive. This is due to the fact that the upgrade of interest rates increases the payments of the depositors more than the payments that are received. Otherwise, if the liability of a bank includes the deposits for more than a year and the loans of fluctuant interest rate replicate monthly, an increase in the interest rates could contribute to the increase of inflows and the bank could be characterized as asset sensitive.

There is a number of different approaches for the measurement of the ratio of net profit to the interest rate changes. The most simple and known method is the calculation of the gap analysis, which is described below.

Another approach is the modeling of the cash flows of various assets and liabilities from the bank balance sheet as an interest rate function. Thus, it is possible to take into account the dependence of the mortgage payment on the interest rates, since constraints to the changes of interest rate payments are introduced to many mortgage products of fluctuant interest rates. Since the assets and liabilities have been modeled, the variability of the bank net profits could be determined. A more recognizable approach is the use of the model in order to simulate the effect of changes on the net profits after one period for a given interest rate change.

Moreover, the spreads among the interest rates with common maturity are possible to change. Specific loans could be related to LIBOR (London InterBank Offered Rate), while others to the interest rates of treasury bills. Another way to encounter these changes to the spreads is to take into account the effect of past changes in interest rates corresponding with specific historical data, a technique which is applied to the methodology of the present research, as described in Chapter 3.

Although many models have been developed for the interest rate risk management, the most usual are the gap analysis and the duration analysis.

More precisely, the maturity gap measure of interest rate risk is concerned with the impact of a change in the interest rate on income or expense. Financial institutions use maturity gap to measure the impact of changes in interest rates on their net interest income (NII).

$$\text{Gap}=\Delta\text{NII}/\Delta\text{r} \qquad (1.12)$$

Where ΔNII is the change of net interest income and Δr is the change of interest rates.

The concept of duration was discovered almost simultaneously by Federich Macaulay (1938) and Sir John Hicks (1939). However, these two researchers had very different objectives. Macaulay's goal was to define a measure by which two bonds with common maturity but divergent payment structure could be compared. In the Macaulay sense, duration measures when the value of the bond is received and is calculated as follows:

$$\text{M}\Delta = \sum_{t=1}^{N} t \frac{[CP_\tau / 1+i)^\tau]}{[(CP_\tau / 1+i)]} \qquad (1.13)$$

where τ is the time of payment completion, *CP* is the interest rate plus the initial capital, *i* is the interest rate and *N* the time duration of the security. The concept of duration is very useful, since it provides a good approach of the sensitivity of the market value of a security to the changes of interest rates:

$$dx = -(dr) * (\text{M}\Delta) \qquad (1.14)$$

where *dx* represents the percentage of the market value of a security, *dr* represents the change to the interest rate and MΔ represents the years .

Hicks attempted to measure interest rate sensitivity for any particular bond. According to Hicks, duration provides a measure of the exposure of the bond to interest rate risk and is defined as follows:

D=-(percentage change in value)/ (percentage change in discount rate)

$$\begin{aligned} D &= -(\Delta V / V) / [\Delta r / (1+r)] = \\ &= -(\Delta V / \Delta r) \times [(1+r) / V] \end{aligned} \qquad (1.15)$$

where *D* is the duration, $\Delta V/V$ is the percentage change in value *V* of the bond, and $\Delta r/(1+r)$ is the percentage change in discount rate r.

The problem of the duration analysis is that it is only a linear measure of risk. Duration is a linear approximation of the value profile of the asset or liability. If the true value profile is nonlinear and if the change in interest rates is large, the approximation error can be large. Consequently, most financial institutions would also measure convexity.

As already mentioned, duration is essentially a measure of the slope of the value profile at current interest rates and bond prices, while convexity is a measure of the curve of the value profile, at current interest rates and bond prices. Thus, the approximation error can be large for large changes in the interest rate, but the approximation error will be larger than that if the risk manager were to look at duration alone.

Recently, simulation models have been developed for the generation of scenarios in the values of interest rates and the determination of the interest rate risk. There is a large number of simulation techniques in the market, which are used for the asset liability management.

The most known simulation technique for the evaluation of these changes is Monte Carlo simulation, which determines how the subject financial variable changes. For example, the most usual assumption is that the interest rates follow a logarithmic probability of normal distribution. Moreover, it defines the initial value and the instability of the financial variable. Based on the above it simulates a number of possible values (from 1000 to 10000) that the financial variable could take during the transaction of the risk management. For each of the simulation values, it calculates the corresponding value to the pre-specified intervals and finally compares the evaluations of the transaction, in order to conclude the distribution probabilities to the pre-specified intervals (Smithson, 1998).

The advantage of the Monte Carlo simulation against the scenario approach is that it takes into account the conjecture of the interest rate changes. More extensive presentation of Monte Carlo simulation is outlined in Chapter 3.

4. THE PROPOSED METHODOLOGICAL APPROACH AND THE OBJECTIVES OF THE BOOK

Changes in the economic environment and specifically the creation of a monetary union led to the development of a financial system in all sectors.

The last decade is characterized by an intense development and change of bank interest rates. The transition to this new status demands the development and implementation of high quality systems for the evaluation of the risks that are undertaken.

Moreover, the asset liability management aims at the optimal balance of the composition of the asset of a bank and of its liabilities, which are connected with the various forms of financial risks, encountered by the bank. The changes of these risks interact significantly with the structure of the assets, liabilities and net profits of the financial institution.

During these changes and taking into account all that were mentioned in the previous sections, it is obvious that the quality of asset management gains great importance as a source of competitive advantage.

Many methodological approaches, stochastic and deterministic, have been proposed for the development of asset liability management models. In the present methodology the data of asset, liability and net income of the financial statements of a bank for the year t are taken into account and a goal programming model is developed for the determination of the assets and liabilities of a bank for the time period $t+1$. A series of goals and constraints, reflecting the demands and preferences of the bank managers as well as the bank policy and strategy, is formulated. The specified goals and constraints concern the structure of the balance sheet, the solvency and liquidity ratios, that is the ratio of equity to the total weighted data of asset and the ratio of quick assets to the short-term liabilities respectively, the average yield of asset and liability, as well as constraints on the variables of fixed assets, cash and funds to Central Bank. Based on these goals and constraints a goal programming system is developed. The solution of this model contributes to the optimization of the deviations from the specified priority levels of the system goals. In order to select the priority levels two alternatives are taken into account, according to which first priority level is given to the solvency goal, second priority level to the liquidity goal and third to the remaining goals. According to the second alternative first priority level is given to the liquidity goal, second to the solvency goal and third to the remaining goals.

Moreover, changes in the interest rates contribute to the variability of the asset and liability composition. It is obvious that the ratio of average yield of asset and liability is directly influenced, since the accounts of asset and liability are related to the interest rate risk and more specifically to the deposit, loan and bond interest rates. In order to confront the uncertainty for the management of the pre-specified goal of average yield, the Monte Carlo simulation model is used for the generation of scenarios in the interest rates. The goal programming model, which is developed, is solved for each of the interest rate scenarios and the efficient solutions are taken as data of the financial statements of the year $t+1$ for the forecasting and determination of the structure of the assets and liabilities of the year $t+2$.

Concluding, the present book aims at the development of an asset liability management model, which provides the possibility to the financial institutions and more specifically to the banks to proceed to various scenarios of their economic progress for the future, aiming at the management of deposit, loan and bond interest rates emerging from the changes of the market variables.

The originality of the present research consists in the use of the goal of average yield of asset and liability and the selection of the deposit, loan and bond interest rates. This selection is based on the fact that the deposit interest rates constitute the most important source of expense for a commercial bank, while the loan interest rates constitute the most important income account. The uncertainty of the changes to the values of the pre-specified interest rates is encountered with the Monte Carlo simulation model for the generation of scenarios in the interest rates. Moreover, the selection of two alternatives for the determination of the priority levels in the goals of the problem and the determination of the structure of the asset and liability for the year t+2 based on the financial data of year t contributes to the originality of the present book. Finally, the contribution of the present volume to the management science and further to the banking field, consists also, except from the above, in the fact that for the first time an asset liability management model combining the goal programming system and the Monte Carlo simulation is developed and applied to a large commercial bank. More specifically, the contribution of goal programming techniques in bank asset liability management is analyzed.

Chapter 2

Review of the asset liability management techniques

1. ASSET LIABILITY MANAGEMENT TECHNIQUES

Asset and liability management models can be deterministic or stochastic (Kosmidou and Zopounidis, 2001). Deterministic models use linear programming, assume particular realizations for random events, and are computationally tractable for large problems. The banking industry has accepted these models as useful normative tools (Cohen and Hammer, 1967). Stochastic models, including the use of chance-constrained programming, dynamic programming, sequential decision theory, and linear programming under uncertainty, present computational difficulties.

1.1 Deterministic models

The first mathematical models in the field of bank management appeared in the early 60's. The deterministic model of Chambers and Charnes (1961) is the pioneer on asset and liability management. Chambers and Charnes were concerned with the formulation, exploitation, interpretation of uses and constructions which may derive from a mathematical programming model which expresses more realistically the actual conditions of current operations rather than the past efforts. Their model corresponds to the problem of determining an optimal portfolio for an individual bank over several time periods in accordance with the requirements laid down by bank examiners which are interpreted as defining limits within which the level of risk associated with the return on the portfolio is an acceptable one.

The authors' goal is a financial model of a bank's portfolio over several time periods as means of studying the implications of Federal Reserve Bank policy. It is assumed that the banker knows the levels that will prevail, on various dates in the future, of demand and time deposits, of rates of interest and of the bank's net worth. The banker aims to the maximization of profit. He has a choice between various kinds of earning asset (loans, government securities, bonds issued by agencies other than the government) and for each kind of asset he also has a choice of several maturities. He is required to observe two restrictions: a reserve requirement which states that part of the bank's assets must be held in cash or deposited with the Federal Reserve; and a less familiar constraint that maintains what we shall call a "balanced" portfolio. By referring to the "balanced" portfolio, Chambers and Charnes mean a set of measures used by the bank examiners of the Federal Reserve System. They have tried to find the most profitable portfolio plan that could be followed by a bank which satisfied the bank examiners at all times. The criteria of the examiners reflect their judgment of what kind of portfolios are reasonably safe, given the uncertainties which banks face. It seems likely that a bank which satisfies these requirements will be in a good position to meet the contingencies of fluctuation in its deposits and changes in market rates of interest, without running much risk of large losses.

In obtaining the solution to our problem, Chambers and Charnes incidentally generate information which can be used to study the implication of Federal Reserve policy and to solve some of the operating problems of their model bank.

Cohen and Hammer (1967), Komar (1971), Robertson (1972), Lifson and Blackman (1973), Fielitz and Loeffler (1979) have applied successfully the Chambers and Charnes' model. Even though these models have differed in their treatment of disaggregation, uncertainty and dynamic considerations, they all have in common the fact that they optimize a single objective profit function subject to the relevant linear constraints.

Cohen and Hammer (1967) first used the term policy constraint in the context of a commercial bank planning model. The term has since been widely used in the literature. Cohen and Hammer describe an important and complex analytical model developed by the Management Science group at Bankers Trust Company. This model has been an operational tool in the asset management process for several years. Thus, those interested in the financial application of analytical techniques can feel better in the fact that at least one situation exists in which complex models are being practically and usefully implemented. The central problem of asset management revolves around the bank's balance sheet. The optimal balance between these factors cannot be found without considering important interactions that exist between the structure of a bank's liabilities and capital and the composition of its assets. The asset management model of Cohen and Hammer concerns a linear programming model. Linear programming can be viewed as a technique for maximizing a linear criterion function subject to a set of linear constraints. The present model includes three stages: intra-period constraints, inter-period considerations and the criterion function.

Fielitz and Loeffler (1979) describe the development of a mathematical programming model for liquidity management in a medium to large commercial bank. Although there are differences in the models, the motivation for the application of this methodology comes from Komar (1971). The application of a mathematical programming model is in part a response to such changes in the nature of financial markets. Moreover, the application is motivated by recognition that, even while liquidity is viewed as supporting other banking functions, it is possible to realize significant returns by actively managing liquidity for profit. Liquidity management is characterized by the need to achieve a formal objective in a constrained environment.

The liquidity management model requires input information reflecting money and capital market supply and demand conditions, tax rates, anticipated credit demands and deposit withdrawals and other factors affecting commercial bank liquidity. The output provided by the model indicates the amount to be held of each type of liquid asset and liability and the highest net earnings consistent with the constraints. The liquid variables are assumed to be continuous, while the relationships among variables are assumed to be linear. The decision variables of the Fielitz and Loeffler model coming from the asset categories are the treasury securities, the agency securities, the municipal securities, the project notes, the federal funds, the certificates of deposit, the repurchase agreements and the Federal Reserve discount window borrowings. The objective function of the model is stated in terms of maximizing the after-tax profit generated from management of the liquidity variables, while the constraints come from the external environment, the internal risk and the return preferences of bank management. More specifically, the constraints refer to restrictions imposed for accounting, legal, regulatory, or market reasons. Such restrictions include the activity level constraints, the pledging constraints and the cash flow constraint. The activity level constraints prevent the model from selling more of an asset than is held. Moreover, commercial banks are required to hold collateral for the acquisition of certain types of liabilities. Certain variables (sources of funds) contribute positively to cash flow and certain variables (used of funds) contribute negatively. Moreover, portfolio composition constraints might limit the amount of funds that could be derived from a particular source, or might affect the distribution of funds among liquid assets. To insure that the bank maintains adequate liquid assets for projected and unanticipated deposit withdrawals or credit demands, a liquidity capability constraint is included in the model. Several formulations of this constraint are possible. In this case, Komar suggests that changes in net liquid assets should be matched with changes in deposits. Fielitz and Loeffler suggest that this constraint provides that net liquid assets should be no less than a given percentage of total assets. To further reflect the subjective evaluation of future economic conditions and to compensate partially for the absence of a multiperiod framework, the liquidity manager may manipulate the values of the maturity constraint. Finally, the securities' gain (loss) constraint is imposed. This constraint is useful to the extent that bank managers and investors consider earnings after securities gains or losses to be important, the formulation of this constraint can contribute significantly to management of the bank's actual, after-tax earnings. Fielitz and Loeffler applied the above mathematical programming model over the data of the Investment Management Division for a period of one year. The model was applied based on the preferences of the banking managers, who at the beginning of each week defined their forecasting and modulated the constraints. Concluding, Fielitz and Loeffler suggest that the benefits of constructing and implementing such a model are not easily measured. Improved communication and coordination in managing liquidity and better education of the staff are not readily quantifiable. There are, however, indications besides its continued use that the introduction of the model has been successful. Finally, the increase in emphasis on profitability has caused the Investment Management Division to be classified in the bank's accounting system as a profit center rather than simply as a source of funds to support other banking functions.

1.1.1 Multiobjective linear programming model

Eatman and Sealey (1979) developed a multiobjective linear programming model for commercial bank balance sheet management. The objectives used in their paper are based on profitability and solvency. The profitability of a bank is measured by its profit function. Since the primary goals of bank managers, other than profitability, are stated in terms of liquidity and risk, measures of liquidity and risk would seem to reflect the bank's solvency objective. There are many measures of liquidity and risk that could be employed, just as there are many measures used by different banks and regulatory authorities. Eatman and Sealey (1979) measured liquidity and risk by the capital-adequacy (CA) ratio and the risk-asset to capital (RA) ratio respectively. The CA ratio is a comprehensive measure of the bank's liquidity and risk because both asset and liability composition are considered when determining the value of the ratio. Since liquidity diminishes and risk increases as the CA ratio increases, banks can maximize liquidity and minimize risk by minimizing the CA ratio. The other objective reflecting the bank's solvency is the RA ratio. Using the RA ratio as a risk measure, the bank is assumed to incur greater risk as the RA ratio increases. Therefore, in order to minimize risk, the RA ratio is minimized. The constraints considered in the model of Eatman and Sealey are policy and managerial. Generally, Eatman and Sealey developed a multiobjective linear programming model and demonstrated its usefulness as a tool for commercial bank balance sheet management. Since bank balance sheet management is clearly a multiple objective decision-making process, optimal utility maximizing solutions require explicit consideration of this multiobjective characteristic. Multiobjective linear programming appears to have a number of advantages over more traditional programming techniques. First, it allows the introduction of all the information required to achieve a utility maximizing solution and gives the opportunity to the decision makers to choose explicitly the final course of action.

Apart from Eatman and Sealey, Giokas and Vassiloglou (1991) developed a multiobjective programming for bank assets and liabilities management. They supported that apart from attempting to maximize revenues, management tries to minimize risks involved in the allocation of the bank's capital, as well as to fulfill other goals of the bank, such as retaining its market share, increasing the size of its deposits and loans, etc. Conventional linear programming is unable to deal with this kind of problem, as it can only handle a single goal in the objective function. Goal programming is the most widely used approach in the field of multiple criteria decision making that enables the decision maker to easily incorporate numerous variations of constraints and goals.

In the Appendix at the end of the present chapter, the linear goal programming model is described.

1.2 Stochastic models

Apart from the deterministic models, several stochastic models have been developed since the 1970's. These models, in their majority, originate from the portfolio selection theory of Markowitz (1952, 1959) and they are known as static mean-variance methods. According to this approach the risk is measured by the variance in a single period planning horizon, the returns are normally distributed and the bank managers use risk-averse utility functions.

Markowitz's model (1952)

More specifically Markowitz presented a construction model of efficient portfolios. The introduction of the meaning of risk and its effect on the decisions of the average investor has brought the revolution to investment decisions till now. According to Markowitz the average investor has two basic objectives, the maximization of the expected return and the minimization of the risk, which is defined as the variance of the return. It supports that as two stocks could be compared examining the expected return and the standard deviation of each of them, the same could occur for both portfolios. The expected return of a portfolio is calculated as the weighted average of the expected returns of the stocks that constitute them. Since the standard deviation or the variance of a portfolio is equal to the covariance of the returns of stocks that constitute them, the risk is reduced as long as the stocks that are involved in the portfolio increase. The negative return of the stock is compensated by the positive return of the other stock and thus there is diversification of the portfolio. According to Markowitz diversification means that a portfolio could not be constituted by the stocks of only one branch, because these stocks have positive correlation between them. Finally, Markowitz's model is based on several conditions such as the efficacy of the financial market. Moreover, the investor has as keeping time a single period. He tries to maximize the return of his capital, minimizing the risk and selects a portfolio, based on the average return of the stocks that constitute them and of the covariance. Besides, the stocks of the portfolio could not have a positive correlation between them and securities without risk are not included to the portfolio. Based on the theory of Markowitz, in order to generate a model, several data, considered as stable, should be isolated and presented as significant. However, the efficiency of a model is determined by its ability to contribute to the procedures that it describes and to produce reliable forecasts for the future.

Pyle's model (1971)

Pyle applied Markowitz's theory in his static model where a bank selects the asset and liability levels it wishes to hold throughout the period. He considers only the risk of the portfolio and not other possible uncertainties. The model omits trading activity, matching assets and liabilities, transaction costs, and other similar features. The essential characteristic of Pyle's study is that it refers to the theory of financial intermediary, who uses the proceeds to purchase other financial assets. The model of financial intermediation developed by Pyle has its roots in the hedging model of Telser (1955-6) and Houthakker (1968). The Telser model is based on the "safety-first" principle of investor choice under uncertainty while Houthakker takes a portfolio selection approach which is in the spirit of the Markowitz model. Pyle shows that these two approaches to choice under uncertainty lead to the same set of solutions to any hedging problem. The principal question on which the analysis of Pyle is based is: under what circumstances would a firm be willing to sell a given deposit liability and use the proceeds to purchase a given type of financial asset. The model that is developed by Pyle deals with the portfolio problem in intermediaries while it neglects the important problems of liquidity and transactions in intermediaries.

Concluding, the literature on the theory of financial intermediation has concentrated on either the asset side or the liability side of the balance sheet. By explicitly considering the dependence between the securities bought and sold by financial intermediaries, it has been shown that asset portfolios cannot, be chosen independently of the parameters of liability yields. The major result of the paper is contained in the specification of the yield relationships which are conductive to financial intermediation. The major result of the study includes the specification of the yield relationships which are conducive to financial intermediation. Of course, there are many unanswered questions regarding the theory of financial intermediation particularly in terms of more dynamic models of the portfolio problem the intermediaries face.

Brodt's model (1978)

A more sophisticated approach was that of Brodt, who adapted Markowitz's theory and presented an efficient dynamic balance sheet management plan that maximizes profits for a given amount of risk over a multiperiod planning horizon. His two-period, linear model included uncertainty and based on the portfolio selection theory of Markowitz. He tried to build the efficient differentiation between the function of expected profits and the linear one of its deviations. Instead of the variance, he used the mean absolute deviation or the semi-absolute deviation that is taken by varying the value of the upper or lower bound of one of the two functions.

1.2.1 Chance constrained programming models

Charnes and Thore (1966), Charnes and Littlechild (1968) developed chance constrained programming models. These models express future deposits and loan repayments as joint, normally distributed random variables, and replace the capital adequacy formula by chance-constraints on meeting withdrawal claims. These approaches lead to a computationally feasible scheme for realistic solutions.

Charnes and Thore's model (1966)

Charnes and Thore present an application of the method of chance constrained programming to a case of financial planning. It is obvious that this new method of programming will be helpful in the future in analyzing a wide range of problems in the field of financial budgeting and the costing of funds for corporations and financial institutions. Indeed, the method of chance-constrained programming was explicitly developed to take care of the following two fundamental aspects of planning, both of which are characteristic of the setting of the problem in financial budgeting: a) the large number of constraints institutional, subjective or others and b) the uncertain future, the chance elements entering both the object function to be optimized and the constraints. The Charnes and Thore study brings an example from the field of financial planning in savings and loan associations. A chance constrained model with linear decision rules is developed and analyzed. Detailed comparisons are made with suggestions in the literature on the optimal management of savings and loan association holdings in the face of liquidity reservations and aversion to excessive debt. It is shown that under certain circumstances these prescriptions are indeed optimal. Among other results, the solved prototype explicitly indicates the circumstances under which the constraints are violated.

Pogue και Bussard's model (1972)

Pogue and Bussard have formulated a 12-period chance constrained model in which the only uncertain quantity is the future cash requirement. Pogue and Bussard were the first ones to formulate the short term financial planning problem under uncertainty as an optimization model. The short term financial planning problem can be structured within a mathematical programming framework, where the objective is to minimize the short run financing costs subject to such constraints. This fact was recognized by Robichek, Teichroew and Jones (RTJ), who developed a model for optimal short term planning under uncertainty. This model takes all of the input data requirements of the model as known with certainty at the beginning of the planning interval. These include the forecasted cash requirements during each subperiod of the planning period. The model developed by Pogue and Bussard is an extension of the RTJ formulation. The extensions to their model are that a) the major source of uncertainty in the problem has been treated. The approach has been reformulated to allow explicit consideration of the uncertainty associated with the forecasted cash requirements. This extension results in the tendency to maintain liquidity to protect against the possibility that future cash requirements may be higher than currently predicted, b) financing options have been generalized to include commercial paper and multiple period investment options, c) the model has been reformulated in terms of stock variables rather than flow variables.

Concluding, the model of Pogue and Bussard permits the financial manager to incorporate his subjective estimates regarding the uncertainty of future cash requirements into the development of the optimal short run financial plan. The liquidity reserve features of the model permit the maintenance of any desired degree of protection against the possibility of not being able to meet future cash requirements from planned sources of funds. The liquidity reserve contains the firm's marketable securities balances plus any unused borrowing potential on immediately available financing sources.

In most applications only the first period decisions would be implemented. The model would run at the end of each month using revised estimates of future cash requirements and financing costs. It is necessary, however, to prepare the complete short term plan since the first period decisions will depend upon expectations about conditions and courses of action in later periods of the planning interval. Generally, the Pogue and Bussard model could be a useful tool in the development of short term financing decisions for any firm with short run planning problems involving alternative periods of cash surpluses and deficits, a variety of financing alternatives and constraints, and explicit requirements for protection against the uncertainties of future cash requirements.

The major weakness is that the chance-constrained procedure cannot handle a differential penalty for either varying magnitudes of constraint violations or different types of constraints.

1.2.2 Sequential decision theoretic approach

In 1969, Wolf proposed the sequential decision theoretic approach that employs sequential decision analysis to find an optimal solution through the use of implicit enumeration. This technique does not find an optimal solution to problems with a time horizon beyond one period, because it is necessary to enumerate all possible portfolio strategies for periods preceding the present decision point in order to guarantee optimality. In order to explain this drawback, Wolf makes the assertion that the solution to a one-period model would be equivalent to a solution provided by solving an n-period model. This approach ignores the problem of synchronizing the maturities of assets and liabilities.

The purpose of the study of Wolf is to formulate a normative model for selecting a bank's government security portfolio. Two major problems arise in constructing a model of bank portfolio selection. First, the model must handle uncertainty. This includes not only uncertain future events but also the decision maker's preferences for the outcomes associated with these events. Second, it must recognize the intertemporal or multi-period character of the decision making process. This means that a decision made in one period will influence subsequent decisions, which must be considered in arriving at the present one. The study of Wolf applies Bayesian and sequential decision theory to handle both the stochastic and dynamic aspects of this important decision problem simultaneously and consistently. No previous model of commercial bank portfolio selection handles either or both problems satisfactorily. Porter's model (1962) of bank asset selection recognizes uncertainty by treating future cash flows and security prices as random variables, but only in one period. Moreover, it does not consider the decision maker's preferences. Since the objective function is linear, the model produces a portfolio diversified between securities and loans only through the selection of distribution functions describing the random variables. Cheng's model (1962) of bank security portfolio selection is also a one period formulation. It incorporates uncertainty and the decision maker's preferences through Markowitz's efficient portfolio concept. An efficient portfolio is one which maximizes expected return for a given variance of return. As Tobin (1958) proves, this criterion assumes that either the variable return is normally distributed or that the decision maker has a quadratic utility function.

Multi-period bank portfolio selection models are all based on the assumption that future events are known with certainty. One such model formulated by Chambers and Charnes attempts to reflect the risk inherent in different portfolio configurations by including the Federal Reserve's capital adequacy formula as a constraint. The capital adequacy formula allocates a bank's capital to designated asset categories on a fractional basis. The values of the fractions are designed to measure the percent by which the different asset categories would decline in market value if they had to be liquidated quickly. Moreover, the formula itself implicitly assumes a particular preference structure and a certain probabilistic occurrence of future events. Neither assumption is likely to represent accurately either the decision maker's preferences or expectations.

Bradley and Crane's model (1972)

Bradley and Crane have developed a stochastic decision tree model that has many of the desirable features essential to an operational bank portfolio model. The Bradley-Crane model depends upon the development of economic scenarios that are intended to include the set of all possible outcomes. The scenarios may be viewed as a tree diagram for which each element (economic condition) in each path has a set of cash flows and interest rates. The problem is formulated as a linear program, whose objective is the maximization of expected terminal wealth of the firm and the constraints refer to the cash flow, the inventory balancing, the capital loss and the class composition. To overcome computational difficulties, they reformulated the asset and liability problem and developed a general programming decomposition algorithm that minimizes the computational difficulties.

The bond portfolio problem is viewed as a multistage problem in which buying, selling and holding decisions are made at successive points in time. Normative models of this decision problem tend to become very large, particularly when its dynamic structure and the uncertainty of future interest rates and cash flows are incorporated in the model. The study of Bradley and Crane presents a multiple period bond portfolio model and suggests a new approach for efficiently solving problems which are large enough to make use of as much information as portfolio managers can reasonably provide. The procedure uses the decomposition algorithm of mathematical programming and an efficient technique developed for solving subproblems of the overall portfolio model. The purpose of the procedure is the definition of subproblems which are easily solved via a simple recursive relationship.

The model primarily focuses upon selection of bank investment portfolio strategies, although it potentially could be expanded to include other bank assets. This concentration on the investment portfolio requires the assumption that portfolio decisions are residual to other bank decisions such as loan policy. The size of the portfolio is therefore determined outside the model framework by the flow of loans, deposits, and borrowed funds. An uncertain amount of funds is made available to the portfolio over time and the optimal investment strategy must take this liquidity need into account. A second external constraint is also imposed upon investment strategy by the bank's available capital. This affects the riskiness of the investment portfolio which the bank is willing to tolerate and is expressed through limits on maximum capital losses. The structure of the portfolio model presented here assumes that decisions are made at the start of each period. At the beginning of the process, the manager starts with a known portfolio and he faces a known set of interest rates. To put it simply it is assumed that the initial portfolio is cash, but this assumption can be easily changed. He can invest in any of a finite number of asset categories which can represent maturity groups and types of bonds, such as U.S. Governments and municipals. The results of this initial investment decision are subject to some uncertainty, represented by a random "event" which occurs during the first period. An event is defined by a set of interest rates and an exogenous cash flow. For example, one event might represent a tightening of credit conditions in which interest rates increase and the portfolio size has to be decreased to finance a rise in loans. It is assumed that there are a finite number of such events which have a discrete probability distribution. In addition, the portfolio manager knows the probability of each event and it is appropriate for him to behave as he did. Due to this assumption, the study of Bradley and Crane treats risk and uncertainty synonymously.

The three-period three-event problem used to illustrate the model structure is perhaps misleadingly small for illustrating the requirements of a realistic bond portfolio problem. Let consider a four-period model, covering two years where the time periods are 3 months, 6 months and 1 year. Variable length periods are chosen to reflect uncertainty about events more distant in the future. Let the security classes represent maturity categories of U.S. Government bills, notes and bonds, 3, 6 months, 1, 2, 3, 5 10 and 20 years. Buying, selling and holding decisions must be made for each of these security classes at the start of each period conditional on the preceding sequence of uncertain events. After the decisions in each period a random event occurs which determines the set of interest rates and the exogenous cash flow for the subsequent decision. The number of random events in each time period depends on the degree of detail desired and the resulting problem size. For the four-period model under consideration a five-point approximation is taken to the distribution of interest rates and exogenous cash flow in each of the three-month periods and a three-point approximation in the six-month and one-year periods. The motivation is that over the first six months some reasonable forecasting of interest rates is currently possible. Since a tightening of credit conditions is normally associated with an increase in rates, an exogenous cash outflow is assumed to occur with a rate increase, representing a need for funds in other parts of the institution. Similarly, no exogenous cash flow is assumed to occur with constant rates and a cash inflow is associated with falling rates.

Concluding, the major advantage of Bradley and Crane's approach is its ability to handle large problems which would require an inordinate number of constraints if they were specified as linear programming under uncertainty models. Further research is needed to improve the ability of portfolio managers to supply the needed assessments for larger models and to study the sensitivity of model solutions to increases in the number of time periods and the number of events per period.

1.2.3 Dynamic programming

Another approach to stochastic modeling is dynamic programming. The approach dates back to the work of Samuelson (1969), Merton (1969, 1990) and others. The main objective of this approach is to form a state space for the driving variables at each time period. Instead of discerning the scenarios, stochastic control perplexes the state space. Either dynamic programming algorithms or finite element algorithms are available for solving the problem.

Merton (1969) in his paper explores two classes of reasons why optimal endowment investment policy and expenditure policy can vary significantly among universities. This is done by relating the present value of the liability payments to the driving economic variables. The analysis suggests that managers and others who judge the prudence and performance of policies by comparisons across institutions should take account of difference in both the mix of activities of the institutions and the capitalized values of their no endowment sources of cash flows.

Eppen and Fama (1971) modeled two and three asset problems. The basic idea is to set up the optimization problem under uncertainty as a stochastic control model using a popular control policy. This model reallocates the portfolio in the end of each period such that the asset proportions meet the specified targets. The continuous sample space is represented via a discrete approximation. The discrete approximation offers a wider range of application and is easy to implement. These models are dynamic and account for the inherent uncertainty of the problem.

Samuelson (1969) formulates and solves a many-period generalization, corresponding to lifetime planning of consumption and investment decisions. For simplicity reasons, he uses in his article an example in order to develop a dynamic stochastic programming model. As an example of topics that can be investigated within the framework of the present model, consider the question of a "business man risk" kind of investment. In the literature of finance, one often reads "Security A should be avoided by widows as too risky, but is highly suitable as a businessman's risk". The response to the above question and all the alternative questions that arise led Samuelson to the development of a dynamic stochastic programming model. Since the businessman is more affluent than the widow and being further removed from the threat of falling below some subsistence level, he has a high propensity to embrace variance for the sake of better yield. Moreover, he can look forward to a high salary in the future and with so high a present discounted value of wealth, it is only prudent for him to put more into common stocks compared to his present tangible wealth, borrowing if necessary for the purpose, or accomplishing the same thing by selecting volatile stocks that widows shun. Besides, being still in the prime of his life, the businessman can "recoup" any present losses in the future. The widow or retired man nearing life's end has no such "second or n^{th} chance". Since the businessman will be investing for many periods, "the law of averages will even out for him" and he can afford to act almost as if he were not subject to diminishing marginal utility.

Samuelson's model takes into account all the above speculations and ends that for isoelastic marginal functions, in the beginning of life someone has similar ability of risk tolerance with the end of life. Moreover, the replacement of present investment losses to the future is relative.

1.2.4 Stochastic linear programming

An alternative approach in considering stochastic models is the stochastic linear programming with simple recourse (SLPSR). This technique explicitly characterizes each realization of the random variables by a constraint with a limited number of possible outcomes and time periods. The general description of the stochastic linear programming model is presented at the Appendix of the present chapter.

Cohen and Thore's model (1970)

Cohen and Thore viewed their one-period model more as a tool for sensitivity analysis than a normative decision tool.

It is well-known that linear programming is a powerful normative approach to the analysis of bank portfolio decisions. Plans and decisions affecting the size and structure of a bank's assets, liabilities and capital accounts have a direct and important impact on its profitability, risk and liquidity. The essence of bank dynamic balance sheet management is undeniably planning over time under conditions of uncertainty. There are two major types of trade-off relationships that must be explored in the planning process. First, the nature and magnitude of the trade-offs which exist between yield, liquidity and risk considerations must be assessed. Second, the short run versus long run implications of various decision alternatives must be evaluated; these arise because the yield, liquidity and risk implications of dynamic balance sheet management actions unfold over a relatively long time horizon. The most successful technique that has actually been developed and implemented in any American commercial bank to help senior executives explore and understand the trade-off relationships inherent in dynamic balance sheet management is the large-scale intertemporal linear programming model. The model determines the sequence of period-by-period balance sheets which will maximize the bank's net return subject to constraints on the bank's maximum exposure to risk, minimum supply of liquidity and a host of other relevant considerations.

The study of Cohen and Thore presents the development of a linear programming model for bank dynamic balance sheet management in a major American commercial bank. One of this model's formal assumptions is that accurate forecasts can be made of future loan demand, deposit levels and interest rates and yields over a several-year planning horizon. Despite this simplification of reality, this model has proved useful in helping executives plan a sequence of period-by-period balance sheets that maximizes the bank's net return subject to constraints on the bank's maximum exposure to risk, minimum supply of liquidity and a host of other relevant considerations. From a practical point of view, the use of sensitivity analysis can help compensate for the model's certainty regarding future economic developments. From a theoretical viewpoint, however, it would be desirable to incorporate explicitly into the programming model, some aspects of uncertainty regarding forecasts of the future. For this purpose, the method of two-stage linear programming under uncertainty with discrete distribution functions seems particularly promising.

A major advantage of this method of programming under uncertainty is its computational appeal. It provides an easy and accurate method of extending a given linear programming model for bank dynamic balance sheet management to incorporate selected uncertainty features, such as uncertainty in future movements of deposits and uncertainty in future loan demand.

This approach to programming under uncertainty using discrete distribution functions would appear to be practical only as long as the number of stochastic outcomes is low. It is not proposed as a general method of programming bank portfolios under uncertainty. It is suggested for use mainly as a type of sensitivity analysis, enabling the analyst to replace selected forecasts with explicit discrete probability distributions, estimating the effects of these uncertainties to the optimal solution. Of particular importance is the potential ability of this type of model to help bank executives obtain insight into the manner in which these uncertainties regarding future economic developments influence the optimal set of first stage, first period decisions. These are the only current decisions that need to be implemented before additional information about the future is obtained, at which time the programming model can be reformulated and run.

Cohen and Thore's model is separated in to two parts or stages. The first stage isolates the controlled decision of the bank and the second the random events that affect the bank from these decisions. The maximization of the objective function of the model is related to the operation profits. It is subject to the specified constraints, which reflect the economic environment of the bank. The Cohen-Thore's model assumes that only the savings deposits and current account deposits are subject to uncertainty. Moreover, the model is determined by the various categories of the balance sheet accounts and the decision variables, as well as the relations among the variables and the accounts. Ascertaining that the banks use a portfolio management model, Cohen and Thore suggest some general extensions of the model. These extensions take the form of extra stochastic parameters, decision periods, accounts and classification of decision, as well as economic relations with the objective function.

Booth's model (1972)

Booth applied this formulation by limiting the number of possible realizations and the number of variables considered, in order to incorporate two time periods. The study of Booth is related to the model of Cohen and Thore and suggests few other modifications in its use.

To accomplish the implementation of the above extensions, the model that Cohen and Thore present is generalized to a multiperiod stochastic programming model with recourse. The model is constructed so that it will maximize the total bank profits subject to certain constraints. The model is linear and contains two decision periods that together span one year. In addition, interest rates, loan demand and the supply of deposits and other liabilities are assumed to be random. Booth's model is presented in three parts: 1) the theoretical structure, 2) the specific parameters and variables, and 3) the objective function and constraints.

Concluding, Booth's model is an extension of the Cohen-Thore portfolio management banking model. First, the Cohen-Thore model allows only deposit levels to be stochastic. Booth's model considers interest rates, loan demands and non deposit levels to be stochastic as well as deposit levels. Moreover, the model developed by Cohen-Thore is a one period model, while this developed by Booth considers two periods. This multi-period characteristic helps against myopic portfolio management decisions. In addition, Booth's model is constructed so that the data required as input is obtainable for most banks. Finally, Booth's model considers important economic phenomena not explicitly present in the Cohen-Thore model.

Nevertheless, even though Booth's model extends the Cohen-Thore model, it could be further developed. One direction is to modify the model's stochastic structure. This modification may involve increasing the number of states of nature from the present three, adding more time periods to the present two, and considering the joint probability distributions of the model's stochastic parameters. Either type of modification, while conceptually uncomplicated, increases the number of constraints and variables in Booth's model in a multiplicative fashion. The resulting number of constraints and variables could easily exceed the limits of current computer software and hardware. In any event, the required input data would be substantially expanded and the model's output would be difficult to interpret. These input and output characteristics would most likely pose an unreasonable burden on the bank. At the present "state of art", this burden may outweigh the benefits derived from this type of modification.

However, there are two extensions that show promise and are feasible. First, the balance sheet accounts may be further disaggregated. This extension only adds a relatively small number of variables. For example, commercial loans could be classified as low risk, medium risk and high risk loans. These different classes of commercial loans would be associated with different interest rates and default characteristics. The extended model of Booth is able to weigh these factors and indicate to the bank what type of commercial loan it should make. Second, in addition to the cost parameter and dividend policy modifications previously mentioned, other constraints could be added to Booth's model, to reflect specific managerial desires.

Booth's model would be useful to banks in the development of portfolio management strategies. The model provides a vehicle for the bank to investigate the effect of alternative economic assumptions on its portfolio management decisions. Since the required input data are relatively convenient to gather, this investigation may be conducted efficiently. Moreover, the model provides the bank with an indication as to which economic assumptions need to be accurately specified.

Nevertheless, it cannot be emphasized that the type of model described in this paper is a policy decision model and not an operating decision model. The function of Booth's model is only to suggest methods by which the bank may achieve its stated financial objectives. These methods must be tempered by banking knowledge and by social objectives not considered by the model.

Crane's model (1971)

Crane on the other hand, modulated the model to a two-period one. The computational tractability and the perceptions of the formulation precluded consideration of problems other than those that were limited both in terms of time periods and in the number of variables and realizations.

Crane presents a discrete stochastic programming model for commercial bank bond portfolio management. It differs from previous bond portfolio models in that it provides an optimization technique that explicitly takes into consideration the dynamic nature of the problem and that incorporates risk by treating future cash flows and interest rates as discrete random variables. The model's data requirements and its computational demands are sufficiently limited so that it can be implemented as a normative aid to bond portfolio management. In addition, it can be extended by the addition of other asset and liability categories to serve as a more general model for commercial bank asset and liability management.

The normative model can also be used, outside the context of a particular bank, as an experimental tool for studying the structure and behavior of optimal bond portfolios. Crane shows that some traditional practices of bank portfolio management frequently may lead to suboptimal portfolio strategies. Moreover, it is shown that the elimination of special capital gains tax treatment for banks will change the maturity structure of optimal portfolios.

The study of Crane deals with the portfolio problems as a multistage decision problem under risk. The primary focus of this study concerns a major aspect of bank asset and liability management, that of selecting investment portfolio strategies. The specific models separate the categories of asset data and select the bonds. This separation of the investment problem involves two major assumptions. First, the maximum size of the portfolio is determined outside the model framework. Second, so long as the portfolio decisions satisfy the cash flow constraints imposed by the rest of the bank, portfolio strategies can be selected without regard to the return or risk associated with other bank assets.

It would be also desirable to avoid these assumptions in a complete model for bank asset and liability management. The assumptions appear reasonable, however, when compared to actual bank practice. Several studies and descriptions of bank management note that the size of the bank security portfolio tends to be treated as a residual after other commitments of the bank are met. For example, the portfolio manager must take into account possible flows in and out of the portfolio stemming from changes in the loan volume of major bank customers. These loan decisions tend to be made independently of their effect on portfolio earnings. To the extent that other management decisions are made in the same manner, it is both convenient and reasonable to formulate a separate portfolio model, as presented by Crane.

The bank portfolio problem can be viewed as a multistage decision problem in which portfolio actions are taken at successive points in time. At each decision period, the portfolio manager has an inventory of bonds and cash on hand. Based upon present credit market conditions and his assessment of future interest rates and cash flows, the manager must decide which bonds to hold in the portfolio over the next time period, which bonds to sell, and which bonds to purchase from the marketplace. These decisions are made subject to a constraint on total portfolio size, which may be larger or smaller than the previous period's constraint depending upon whether a cash inflow or outflow occurred. At the next decision period, the portfolio manager faces a new set of interest rates and a new portfolio size constraint. He must make another set of portfolio decisions that take into account the new information. This decision-making process is repeated over many time periods.

In order to deal with this decision problem, a normative bond portfolio model should explicitly consider its dynamic multistage characteristics and the uncertainty or risk in future interest rates and cash flows. Thus, in choosing the optimal first-period portfolio, the model should take into consideration the nature and probability of future outcomes and also the optimal actions that would be taken for each of these events. Moreover, it is important for the model to have a technique for obtaining optimal solutions if the model is to be of practical value. Unfortunately previous bond portfolio models have not completely satisfied these criteria. The model of Crane is restricted for expository convenience to a two-stage decision problem. It is assumed that the portfolio manager starts with a known volume of cash of seven assets which can represent maturity categories and types of bonds. At the end of the first period, six months, a random event occurs and the portfolio manager makes another set of decisions in light of the new conditions. An "event" is defined by a set of interest rates and a cash flow that imposes a new portfolio size constraint. There are a finite number of events, three in this particular formulation, that have a discrete probability distribution. Let assume that the portfolio manager knows the probability of each event. At the end of the second six-month period, a new random event occurs and the portfolio is then sold for cash. Thus, the model covers two six-month time periods, with portfolio decisions made at the start of each period and the final value of the portfolio determined at the end of the second period.

Kallberg, White and Ziemba model (1982)

Kallberg et al. have formulated a firm's short term financial planning problem as a stochastic linear programming with simple recourse model where forecasted cash requirements are discrete random variables. The main goal of their paper was to minimize costs of the various sources of funds employed plus the expected penalty costs due to the constraint violations over the four quarter horizon. They concluded that even with symmetric penalty costs and distributions the mean model is significantly inferior to the stochastic linear programming formulation.

The study framework of Kallberg et al. allows a more realistic representation of the uncertainties fundamental to this problem than previous models. Moreover, using Wets (1983) algorithm for linear simple recourse problems, this formulation has approximately the same computational complexity as the mean approximation. Using this formulation, they investigated the effects of differing distributions and penalty costs. They concluded that even with symmetric penalty costs and distributions the mean model is significantly inferior to the stochastic linear programming formulation. Thus, they prove that ignoring the stochastic components in linear programming formulations can be very costly without having significant computational savings.

Their model includes a number of attractive features. In particular, forecasted cash requirements, liquidation and termination costs are all random variables, although for simplicity, in the actual calculations presented, these costs were taken to be deterministic. The objective is to minimize costs of the various sources of funds employed plus the expected penalty costs due to the constraint violations over the four quarter horizon. The data used were adapted from those in Pogue and Bussard (1972). The model is not dynamic since all asset levels are chosen in the initial period. However, it has partial dynamic aspects through the penalties for constraint violations in the fourth trimesters of the year. The solution of the model is very efficient using Wets algorithm for finitely distributed stochastic linear programming problems with simple recourse.

Kusy and Ziemba model (1986)

Kusy and Ziemba employed a multiperiod stochastic linear program with simple recourse in order to model the management of assets and liabilities in banking while maintaining computational feasibility. The model tends to maximize the net present value of bank profits minus the expected penalty costs for infeasibility and includes the essential institutional, legal, financial and bank-related policy considerations and their uncertainties. It was developed for the Vancouver City Savings Credit Union for a 5-year planning period. The model uses initially historical data of past financial statements, takes into account the preferences and goals of the banking managers and solves the problem presenting a series of future forecasts. The objective is the maximization of the bank profits of the next five years based on the following constraints:

1. Deposit flows, where the net cash flow is taken into account during the accounting period, since the deposits are diversified proportionally to the changes of interest rates.
2. Liquidity and solvency, according to which the market value of the bank asset should be enough in order to cover the deposit defaults and not less than the liquidity reserves and liabilities.
3. Policy constraints, which include the relationship between the penalty costs and the extent of policy violations, that is personal loans should not exceed 20% of the first mortgage loans in any period and secondly mortgages should not exceed 12.5% of first mortgages.
4. Legal constraints, as they are determined by government regulations, that is the current assets cannot be less than 10% of the total liabilities (as defined by Credit Union Act – British Columbia Government).
5. Budget constraint, according to which the uses and sources of funds are equal for each period, as defined by the government regulations.

Assuming different probability distributions for the satisfaction of the first three constraints (constraints 4 and 5 remain invariable, if they are defined by government regulations), in proportion to the demands of the banking managers, different optimal solutions are obtained for the bank profits. The optimal solution leads to the achievement of the desirable profit for the next five years. The results indicate that ALM is theoretically and operationally superior to a corresponding deterministic linear programming model and that the effort required for the implementation of ALM and its computational requirements are comparable to those of the deterministic model. Moreover, the qualitative and quantitative characteristics of the solutions are sensitive to the model's stochastic elements, such as the asymmetry of cash flow distributions. Generally, this model is characterized by:

- Multiperiodicity incorporating changing yield spreads across time, transaction costs associated with selling assets prior to maturity, and the synchronization of cash flows across time by matching maturity of assets with expected cash outflows.
- Simultaneous considerations of assets and liabilities to satisfy accounting principles and match the liquidity of assets and liabilities.
- Transaction costs incorporating brokerage fees and other expenses incurred in buying and selling securities.
- Uncertainty of cash flows incorporating the uncertainty inherent to the depositors' withdrawal claims and deposits that ensures that the asset portfolio gives the bank the capacity to meet these claims.
- The incorporation of uncertain interest rates into the decision-making process to avoid lending and borrowing decisions that may ultimately be detrimental to the financial well-being of the bank.
- Legal and policy constraints appropriate to the bank's operating environment.

The Kusy and Ziemba model did not contain end effects, nor was it truly dynamic since it solved two periods at a time in rolling fashion. The scenarios referred to high, low, and average returns that were independent over time.

Russell-Yasuda Kasai model (Carino et al., 1994)

Another application of the multistage stochastic programming is Russell-Yasuda Kasai model (Carino et al., 1994), which aims at maximizing the long term wealth of the firm while producing high income returns. This model builds on this previous research to make a large scale dynamic model with possibly dependent scenarios, end effects, and all the relevant institutional and policy constraints of Yasuda Kasai's business enterprise. The multistage stochastic linear program, used by Carino et al., incorporates Yasuda Kasai's asset and liability mix over a five-year horizon followed by an infinite horizon steady-state end-effects period. The objective is to maximize expected long-run profits less expected penalty costs from constraint violations over the infinite horizon. The constraints represent the institutions, cash flow, legal, tax and other limitations on the asset and liability mix over time.

Based on one or more decision rules, it is possible to create an ALM model for optimizing the setting of decision rules or even to create a scenario analysis. These optimization problems are relatively small, but they often result in non-convex models and it is difficult to identify the global optimal solution. Examples of optimizing decision rules are Falcon Asset Liability Management (Mulvey, Correnti and Lummis, 1997) and Towers Perrin's Opt: Link System (Mulvey, 1996). In general, scenario analysis is defined as a single deterministic realization of all uncertainties over the planning horizon. The process constructs, mainly, scenarios that represent the universe of possible outcomes (Glynn and Iglehart, 1989; Dantzig and Infanger, 1993). The main idea is the construction of a representative set of scenarios that are both optimistic and pessimistic within a risk analysis framework. Such an effort was undertaken by Towers Perrin (Mulvey, 1996), one of the largest actuarial firms in the world, who employs a capital market scenario generation system, called CAP: Link. This was done in order to help its clients to understand the risks and opportunities relating to capital market investments. The system produces a representative set of individual simulations – typically 500-1000, starting with the interest rate component. Towers Perrin employs a version of the Brennan and Schwartz (1982) two factor interest rate model. The other submodels are driven by the interest rates and other economic factors. Towers Perrin has implemented the system in over 14 countries in Europe, Asia and North America.

1.2.5 Simulation models

Derwa (1973), Robinson (1973) and Grubmann (1987) report successful implementations of simulation models developed for various financial institutions. Derwa, for example, used a computer model, operating at "Société Générale de Banque", to improve management decision making in banks. The model was conceived as a form of a decision tree which made it possible to proceed step by step and examine the factors converging on the essential objective of the bank. Derwa concludes that the problems raised by introducing models into management are much more difficult to solve than the technical ones connected with mathematics or date processing. More analytically Derwa's simulation model has the following characteristics:

Deposit and Credit Forecasting Model

In banking, deposits constitute the raw material of the institution. Therefore, it is important to forecast the manner and extent of deposit expansion so that any anomaly in their growth can be identified at the earliest possible moment. The principle followed is to distinguish between the permanent component of the time series and the seasonal and accidental variations.

The values of the variables are estimated by an exponential smoothing stochastic method, in which the smoothing coefficients are such as to give a minimum value to the sum of the squares of the forecasting errors. Using the smoothing coefficients, it is possible to estimate the values of the series for a given number of periods by reference to the last available observation. This method applied to deposits makes it possible to forecast the figure for several months in advance, with an error factor of less than 1%. Knowledge of the seasonal coefficients makes it possible to calculate the most probable course for the target deposit figure. A calculated confidence interval above and below the course figure, indicates the probability of reaching the target by reference to the volume of deposits already brought in. In processing the data, the model is handled on the 8K IBM 1130. The computer is assisted by a Benson Plotter. The system is fully integrated. Past observations are retained on the disk, comprising the amount of each of the different types of deposit at different branches. The information given each month consists only of the last observation available for each series, thus limiting the risk of error and the need for additional checking. The program calculates the forecasts and automatically draws the graph embodying the results. The credit forecasting model, which is operating in the bank, is based on techniques similar to those utilized for the deposit forecasting model Benson Plotter.

Interest rates

Another important element in the management of a bank is an understanding of the mechanism by which interest rates are determined and, if possible, the ability to forecast these rates. The explanatory variables used in this model are conjunctural and monetary. The conjunctural variables are strongly correlated and cannot be introduced as they presently exist into a multiple regression model. In order to avoid the multicollinear effect, they are dealt with factor analysis, and the first principal component is used as the general conjuncture index. The next item calculated is the time lag between the rate of interest and the explanatory variable (the conjunctural index and the monetary variables) which gives the best correlation. For variable for which there is no time lag, the future values are calculated by the model. A stepwise regression, taking into account the optimum time lags and the smoothed values, makes it possible to estimate future values of interest rates. The choice of variables is confirmed by the fact that the step-wise procedure brings in all the series used in the explanatory model.

Financial engineering

Sigma is a linear programming financial engineering model, which is designed to solve the following problem: A firm has decided upon an investment plan, and the question is how to finance the investment most advantageously for the company, taking into consideration the various legal and financial constraints involved. It is also useful for the lending officer, who is better able to appraise the health of a company when it applies for credit.

The financial engineering models that are operational are simulation models. Sigma is an optimization model that checks in advance the compatibility of the assumptions regarding the firm's growth. From the financial viewpoint, Sigma uses the theory of opportunity costs. The objective function to be minimized is the sum of the finance sources weighted in terms of opportunity costs. A unique feature of the system is that the equations of the linear model are introduced as data. This facilitates the analysts to describe the financial structure and policy of the firms to be studied.

Sigma is a multiperiod model. Beginning with the equations for the first year, through the constants brought in for next years, the complete model input is generated in a Report Generator. Using this Report Generator, it is possible to print the results in the familiar form of balance sheets, profit and loss statements and financial tables. Various programs make it possible to modify and process this collection of data.

Sigma also makes it possible to test the coherence of the component items of the firm's finance policy and in case of incompatibility, to indicate the causes. Provided the policy is practicable, Sigma gives a projection of the future balance sheet and profit and loss accounts of the firm, assuming it carries out the optimum financing policy. The model also facilitates a study of the degree to which the recommended solution may be affected by variations in the constants of the system, such as the cost of resources, the amounts to be invested and similar items. The analysts have the necessary elements for a detailed exposition of the consequences of any modification in the financial policy or in the initial assumptions.

Cash balances

Cash balances held in branches of the bank represent a sizable amount of unproductive capital. It is in the interest of the bank to reduce such amounts to as low a level as is practical and possible. The problem is part of the general class of stock problem. The question is one of determining the amounts to be delivered and the release dates that will minimize the total cost, which is the weighted sum of the costs of storage, supply and shortage. There is no way of knowing with any high degree of certainty how much in cash balances a bank branch will need. Nevertheless, it is known that the cash requirements follow a symmetrical distribution and the parameters of this distribution vary from one branch to another and for any individual branch, between one day of the week and another. The model is required to calculate not only the amounts to be delivered but also the frequency of supply. Having calculated all possible transport policies for the week, the one chosen is that showing the smallest total cost.

In order to make the model more realistic, a number of other factors are taken into account. Among these are the denominations of notes and the shorting for bundling, the number of counter positions per branch and such factors as end-of-month or quarterly settlements. The validity of the method was tested by simulation. It was found that through use of the model it was possible to reduce the cash holding by more than 40% without increasing risk of shortage. There are various calculated ratios produced by which performance is measured and the various branch offices compared.

Portfolio Management Model

The object of security portfolio management is to obtain a sufficient return without undue risk. In order to maintain the total return from a portfolio, its composition must be limited to those securities on which the return is highest. In most cases, however, the movements in the prices of these securities are strongly correlated, so that the risk in such a portfolio is high. In case other things are equal, a higher degree of diversification tends to decrease the risk.

This portfolio management model is a Markowitz quadratic programming model. The Markowitz model determines the mix of issues comprising the "efficient" portfolios. Efficient portfolios refer to those portfolios which, for a given return, comprise the smallest degree of risk. The components of the return are the changes in prices and in the dividend or interest income received. The components of the risk are the uncertainty of forecasting relating to each security and the high or low degree of co-variation of the securities among themselves.

From a mathematical viewpoint, minimizing the risk and therefore the total variance of the portfolio is a problem of quadratic programming. Since the aim of the model is to calculate the whole boundary line of efficient portfolios, the Markowitz algorithm is a parametric quadratic program. The variance - covariance matrix between all the shares used in its composition is time consuming to set up.

The program input is of the "Sharpe" variety. The future values of the shares are estimated in the form of a linear relationship with one or more financial indices that may be linked with one another. The correlation matrix is estimated from these relationships, but the possibility exists to feed into the program correlations that have been explicitly calculated elsewhere. The program makes it possible for securities to be classes in groups with an indication of the maximum amount per group and per security. In addition, many types of constraints or limitations are accepted, provided they are linear.

The program output is of a two-fold nature. First, it provides a list of efficient portfolios for each return specified in the program, indicating the risk and the percentage composition. Secondly, for each iteration, a series of technical data can be obtained, making it possible to follow the development of the algorithm. A particular result of these data is the possibility of calculating the dual cost of each constraint, so that the manager in charge can see how much it costs him to depart from the optimum by adding to the problem the task of satisfying any specific condition. From out point of view, these data are the most useful. They supply indications, expressed in terms of budgetary cost, of the relationship to the policy followed in the management of the portfolio. Using the Sharpe method requires forecasting the evolution of the general indices of the stock exchanges.

Arbitrage model

The arbitrage service in a bank is involved in trading of foreign currencies. If foreign exchange markets were perfect, the price of a given currency would be identical in all countries. In practice, however, information is no available instantaneously, and there are a number of distorting factors which create price differences. The purpose of arbitrage is to take advantage of the small differences in price that exist, while at the same time, contributing to the leveling of markets. The great number of data items and their combinations and the frequent price variations, indicate the advantage of using a computer.

The purpose of this dynamic programming model is to take account of the data available and indicate in each case the best way of carrying out an arbitrage operation by sale or purchase, or by borrowing or lending foreign currencies. The model deals with indirect ways of carrying out a given operation. For example, a sale of sterling against deutsche marks may, under certain circumstances, be carried out most advantageously through the French franc.

The model also identifies cycles, i.e. to exchange one currency for another, convert it into a third currency and so on. The imperfections of the market are such that cycles of this type are often available.

The repeated calculation of all the combinations is avoided by using a dynamic programming algorithm. It thus makes it possible to calculate only the simplest optimal paths and to combine them in a complex cycle procedure. Each of the arbitrage operators at the bank has a display unit, which is useful for interrogating the model.

Profit and Loss Simulation

The profit and loss account simulation model is the closest analytical representation of the bank in its economic and commercial environment. It is based on existing accounting diagrams which it sets out in the form of ratios, comparing in pairs the various items in the balance sheet and profit and loss accounts.

The system facilitates a quick determination of the results that may flow from a policy decision by management, or from external changes, such as business conditions, monetary policy or wage settlements. It provides a projection of the balance sheet, profit and loss accounts and cash flow, resulting from each set of assumptions, or from a change in any of the established data. The system can be used for short-term purposes in management and budgetary forecasting. It also helps in the periodic appraisal of target attainment, and in the diagnosis of the causes of discrepancies between target and achievement. On the longer term the model makes it possible to study the consequences of various policies and medium-term strategies in the light of possible changes in external conditions.

From the technical point of view, the model is conceived as a form of decision tree which makes it possible to proceed step by step and examine the chief factors converging on the essential objectives of the bank. It is thus possible to trace the chain reaction to a change in one of the ratios for a district head office of the bank, whether the change is deliberate or involuntary. The consequences of such a decision influence the achievement of the targets of the district office concerned and ultimately the situation of the bank itself.

Management Information System

In running a bank, two different levels of activity can be identified: The conduct of current operations and medium and long-term administration. It is the second level – objectives, resource allocation and control – that constitutes management.

At the operational level, decisions can usually be programmed and the necessary data are easier to determine. This is the reason why computers are primarily used at the operational level. The management information system is involved with attempts to use the computer to facilitate and systemize the work of the senior management in business firms. The further we move away from the operational level, the more difficult it is to draw up systematic rules for decision-making. The questions that must be taken into account are always more numerous and the direct or indirect consequences of decisions are more difficult to forecast.

The establishment of a management information system presupposes: 1) setting up a basis of important data that faithfully represents the firm in its economic, competitive and commercial backgrounds, and 2) establishing a model for the firm, facilitating decision-making at each level and showing the consequences of possible policy decisions of management or changes in the environment.

Concluding, the psychological problems raised by introducing models into management are much more difficult to solve than the technical ones connected with mathematics or data processing.

1.2.6 Dynamic generalized networks

Another approach is the dynamic generalized networks, which was used by Mulvey and Vladimirou (1992) for financial planning problems under uncertainty. They developed a model in the framework of multiscenario generalized network that captures essential features of various discrete time financial decision problems and represented the uncertainty by a set of discrete scenarios of the uncertain quantities. However, these models have a small size and are not able to solve practical sized problems. Mulvey and Crowder (1979) and Dantzig and Glynn (1990) used the methods of sampling and cluster analysis respectively to limit the required number of scenarios to capture uncertainty and maintain computational tractability of the resulting stochastic programs.

Several financial planning problems are posed as dynamic generalized network models with stochastic parameters. These could be the asset allocation for portfolio selection, the international cash management, and the programmed-trading arbitrage. Despite the large size of the resulting stochastic programs, the network structure can be exploited within the solution strategy giving rise to efficient implementations.

Financial planning problems concern the positioning of funds to achieve specific goals. In allocating financial resources, the available options must be analyzed from various perspectives. First, the potential earnings of alternative fund positions must be evaluated in conjunction with the costs associated with the transfers of funds. Second, consumption needs must be met and anticipated future deposits and liabilities must be taken into account. A multiperiod model becomes necessary.

Consideration of uncertainties is critical in financial planning. Major uncertainties include the returns of investment instruments, future borrowing rates and external deposit/withdrawal streams. Investors often seek to develop long-term strategies that hedge against uncertainties.

APPENDIX

ASSET LIABILITY MANAGEMENT PROGRAMMING MODELS

In this appendix the basic equations of the goal programming model and the stochastic linear model, that are used in the aforementioned methodology and indicate the alternative ways of asset liability management, are presented.

1. Linear programming formulation of asset liability management

X_i: Mean balance of asset i

Y_j: Mean balance of liability j

Assuming that r_i is the unit revenue of asset i (in real terms) and c_j is the unit cost of liability j, the objective function is:

$$\text{Maximize } Z = \sum r_i X_i - \sum c_j Y_j \tag{2.1}$$

where Z is the difference between the bank's interest income and the bank's interest expense, that is its revenues beyond operational expenses. The objective function is maximized under a specific set of constraints.

2. Linear goal programming model

The problem of bank asset liability management can be formulated as the following goal programming model:

$$\text{Determine } X = \left(x_1, x_2, \ldots, x_j, \ldots, x_n\right) \tag{2.2}$$

That minimizes the objective function $Z = f\left(d_i^+, d_i^-\right)$

Under the constraints:

$$\sum_{j=1}^{n} c_{mj} x_j \leq \theta_m \quad \text{for } m=1,\ldots,M \tag{2.3}$$

and the goals:

$$\sum_{j=1}^{n} a_{ij} x_j = b_i + d_i^+ - d_i^- \quad \text{for } i=1,\ldots,I \tag{2.4}$$

$$x_j, d_i^+, d_i^- \geq 0 \tag{2.5}$$

where

x_j is the mean balance of asset or liability j (structural variables)

a_{ij} is the consumption coefficient corresponding to x_j in constraint i

θ_m is the target value of each constraint

The constraints (2.3) reflect the availability limitations of resources m and correspond to the constraints in the conventional linear programming model. The goals (2.4) represent the objectives set by management, with the right hand side of each goal consisting of the target value b_i and the positive / negative deviation d_i^+, d_i^- from it.

The difference in formulation between constraints and goals can be handled in a number of ways. In the sequential linear goal programming model which was applied at the Bank, these constraints are transformed to the same form as the goals.

Thus, (2.3) becomes:

$$\sum_{j=1}^{n} c_{mj} x_j = \theta_m + d_m^+ - d_m^- \quad \text{for } m=1,....,M \tag{2.6}$$

The objective function has the following form:

$$Minimize\ Z = \left\{ \begin{array}{l} P_1\left[\sum_{m=1}^{M} W_{1m}(d_m^+, d_m^-)\right], P_2\left[\sum_{i=1}^{I} W_{2i}(d_i^+, d_i^-\right],..., \\ P_\varphi\left[\sum_{i=1}^{I} W_{\varphi i}(d_i^+, d_i^-)\right] \end{array} \right\} \tag{2.7}$$

where

P_φ are the priority levels, with $P_1 \geq P_2 \geq P_\varphi \geq P_{\varphi+1}$

$W_{\varphi i}$ is the linear weighting function of the deviation variables of constraint i at priority level φ,

$\varphi \leq I+1$, i.e. the number of priority levels is less than or equal to the number of goals plus 1, since all constraints appear at the first priority level.

3. Stochastic Linear Program with simple recourse

The general formulation of the stochastic linear programming with simple recourse (SLPSR) model is as follows:

$$\max \quad z = cx - E_{\xi}\{\min(q^{+}y^{+} + q^{-}y^{-})\} \tag{2.8}$$

subject to:

$$Gx = b, \tag{2.9}$$

$$Ax + Iy^{+} - Iy^{-} = \xi \tag{2.10}$$

$$x \geq 0, \quad y^{+} \geq 0, \quad y^{-} \geq 0 \tag{2.11}$$

where $c, x \in R^{n}, y^{+}, y^{-}, q^{+}, q^{-} \in R^{m}, G = m_1 xn, A = m_2 xn,$

x is a decision variable for the asset element, y^{+} and y^{-} is the surplus or deficit of the assets respectively, q^{-} and q^{+} are the penalty rates for the potential withdrawal of funds that are imposed based on the surplus and the deficit, and ξ is a discrete random variable independently distributed by x in the probability interval (Ω, F, μ) (Wets, 1966, 1983). This model can be interpreted as a two-period model, that chooses the vector x without knowing the level of the random vector ξ, observing ξ and making the appropriate adjustments (y^{+}, y^{-}).

The n-period general stochastic linear programming model is described below, where ξ is the discrete random variable with an infinite number of possible solutions (Kusy & Ziemba, 1986, Kira & Kusy, 1990 και Oguzsoy και Güven, 1997).

$$\max z = \sum_{k} c_k x_k - \sum_{i}\sum_{j} p_{ij}\left(q_{ij}^{+} y_{ij}^{+} + q_{ij}^{-} y_{ij}^{-}\right) \tag{2.12}$$

subject to:

$$\sum_{k} g_{hk} x_k = b_h, \forall h \tag{2.13}$$

$$\sum_{k} a_{ik} x_k + y_{ij}^{+} - y_{ij}^{-} = \xi_{ij}, \forall i, j = 1,\ldots,J \tag{2.14}$$

$$x_k \geq 0, \forall k, y_{ij}^{+} \geq 0, y_{ij}^{-} \geq 0, \forall i, j = 1,\ldots,J \tag{2.15}$$

where y_{ij}^{+}, y_{ij}^{-} denote the deficit and surplus of the assets and liabilities, q_{ij}^{+}, q_{ij}^{-} the penalty rates that are imposed regarding the deficit and surplus, p_{ij} the probability of realization of the liability j from the asset i and ξ_{ij} the discrete random variable. Similarly, g_{hk}, a_{ik} and c_k are the coefficients of the k variable x_k to the h deterministic, i stochastic constraint and the objective function (Ziemba, 1974). The application of various methods (Wets, 1983, Birge and Louveaux, 1988, Bryson and Gass, 1994) to the solution of the stochastic linear programming models (SLPSR) contributes to additional efficiency. The equation (2.12) is solved using the approach that meets the dynamical aspects of the n-period problem. In this approach, at the first stage, the vectors are selected by the managers of the variables $x_1, \ldots, x_n$ for the period $1, \ldots, n$ as well as the discrete random vectors $\xi_1, \ldots, \xi_n$ for the same period. At the second stage the variables $\{(y_1^+, y_1^-), (y_2^+, y_2^-), \ldots, (y_n^+, y_n^-)\}$ for the period $1, \ldots, n$ are determined based on the values of the vectors $x=\{x_1, \ldots, x_n\}$ και $\xi=\{\xi_1, \ldots, \xi_n\}$. This approach includes a few dynamic characteristics since the penalty costs of the periods 2 to n are taken into consideration for the determination of the variables $x_1, \ldots, x_n$. The success of this approach is based on the data and the re-iteration, because as it is obvious a model is of the same quality as the data used.

Chapter 3

Bank asset liability management methodology

1. OBJECTIVE OF THE RESEARCH

The review of the asset liability management techniques that were presented in the previous chapter, contributes to the best comprehension of the methodologies that were developed in the field of ALM.

More specifically, during the last years many models concerning the financial analysis were developed. Such models are included in the researches of Kvanli (1980), Lee and Lero (1973), Lee and Chesser (1980), Baston (1989), Sharma et al. (1995), in which goal programming techniques are used in the fields of financial planning and portfolio selection. These studies are focused on the fields of banking and financial institutions and use data from the financial statements of the banks.

The changes, however, and the vulnerability of the market variables necessitate the creation of an ALM system for each financial institution, which provides it with the possibility of evaluating its present financial situation and proceeding to various scenarios of its future economic progress.

The purpose of the methodology that is developed in the present study is the development of a goal programming model in an uncertain environment of changing interest rates. More specifically, the methodology of the present study is presented as follows:

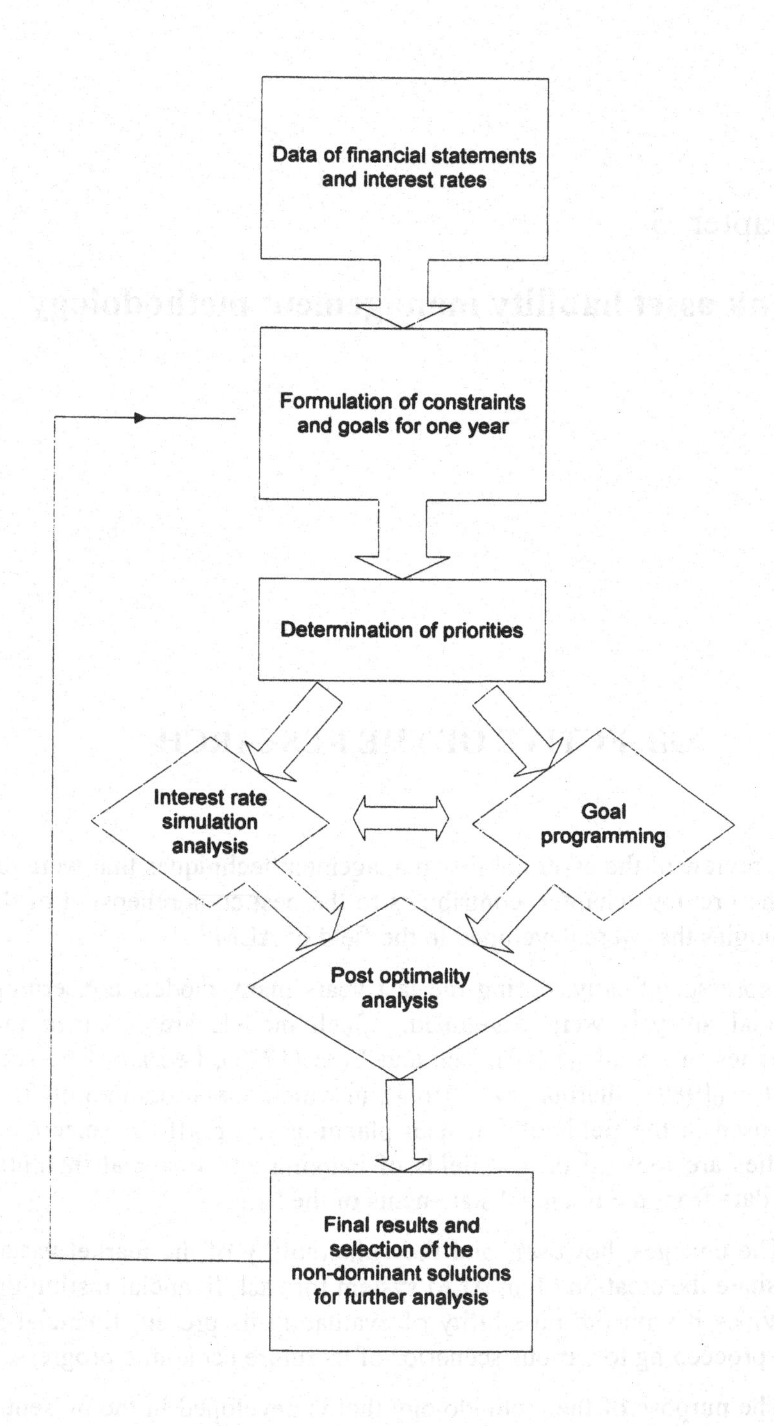

Figure 3.1: Flowchart of the proposed bank asset liability management methodology

Taking into consideration the data of the bank's financial statements for the economic year *t*, we select the accounts of assets, liabilities as well as the total net profits from the income statement for the formulation of the goal programming problem, that aims at the determination of the asset and liability data of the year *t+1*, based on the demands of the bank managers and the bank policy.

Based on the above data, certain constraints and goals that reflect the bank's strategy, as well as the discreteness of the environment in which it works, are developed. Environmental constraints are taken into account. These constraints concern the levels of the deposit and loan accounts, which compose the bank's balance sheet. Moreover, constraints that concern the bank's policy and are imposed on financial ratios, as well as on variables such as the capital, the reserves and the ratio of net profit to total assets, are also taken into consideration. Besides, the structural constraint of the balance sheet that equates the sources to the uses of funds, the solvency goal, the liquidity goal, the goal for the overall expected return of the selected asset liability strategy as well as goals that represent variables such as cash, cheques receivable, deposits at the Central Bank and fixed assets have an essential role in the development of the model. The assigned constraints and goals are reflected in an optimization problem, which has the form of a goal programming problem and aims at the determination of an efficient solution, which maximizes the decision maker's utility. In the framework of the goal programming techniques, each objective function of the problem is converted to a constraint, with the corresponding right hand side representing the ideal value of the goal. Thus, during the development of the goal programming problem, two alternatives that concern the priority criteria of the problem's goals are selected.

Moreover, the interest rate risk indicates the effect of changes to the net profit margin between the borrowing and deposit securities, which develop as a consequence of the deviations of the predominant interest rates. The goal for the overall return of the asset and liability includes the interest rate risk and more specifically the interest rates of deposits, loans and bonds. It is obvious that the changes and fluctuations of the interest rates contribute to the variability of the assets and liabilities, and more specifically, of the goal for the overall expected return. In order to model these parameters various techniques have been proposed (Chapter 1). In the present study, a Monte Carlo simulation is employed for scenario analysis of interest rates, in order to take into account the interest rate risk. The goal programming model is solved for each interest rates scenario. The obtained solutions are then imposed on a dominance analysis for the selection of the efficient optimal solutions.

Then the selected efficient solutions are introduced in the data input (flowchart) and the ALM model is solved in order to determine the assets and liabilities of the next year *t+2*.

This chapter presents analytically the steps of the development of the methodology above.

2. DATA

The application of the asset liability management methodology took place at a commercial bank of Greece. The variables used in the specification of the model, were taken directly from the financial statements and the income statement of the bank for the year *t* in order to choose strategic direction in the bank's financial plan for the year *t+1*.

Moreover, since the variability of the interest rates contributes to the measure of the interest rate risk management and has a direct relation to the change of the assets, the hedging of these risks becomes essential. The interest rates of loans, deposits and bonds are taken into account, in order to proceed to the scenario analysis. A more thorough analysis of the data of the present application is presented in the next chapter.

3. MULTIOBJECTIVE LINEAR PROGRAMMING

Multiobjective linear programming constitutes a generalization of the linear programming and is characterized by the existence of multiple objectives that are under maximization or minimization.

Multiobjective linear programming has been developed during the 70's, and later in the framework of the multicriteria analysis, constituting a philosophy that prevailed as a realistic framework of modeling decision making problems with multiple criteria. Taking into account that either the outcome of a decision cannot be defined exactly (many decision makers, conditions of competitiveness, etc.) or is subject to differentiations during the procedure of decision support, the multiobjective programming enhanced the philosophy of the classical linear programming with new outcomes.

In general a multiobjective programming problem can be described as follows:

$$\max\{g_1(x), g_2(x), \ldots, g_n(x)\} \tag{3.1}$$

subject to:

$$x \in F = \{x \in R^l \,/\, Ax \le b, x \ge 0\} \tag{3.2}$$

where

F is the set of feasible solutions, defined through a system of linear inequalities,

$g_i(x), \quad i = 1,2,\ldots,n$ are the objective functions of the problem, each defined as $g_i(x) = \sum_{j=1}^{l} c_{ij} x_j$, where c_{ij}, $i = 1,2,\ldots,n$, $j = 1,2,\ldots,l$ are the coefficients of the decision variables $x_1, x_2, \ldots, x_l$ in the objective function g_i

A is a $m \times l$ matrix with the coefficients of the decision variables in the m linear constraints

and b is a $m \times 1$ vector with the real coefficients of the constraints.

It should be mentioned, that the search of optimal solution to a multiobjective system, that is the solution that optimizes simultaneously all the functions-goals, is futile, since the criteria have usually a competitive part, so that the optimal as far as a criterion is not the optimal as far as all the others. The methods that have been developed for these systems, aim at the identification of a satisfactory solution that satisfies the decision maker. The radical evolution of computer science has contributed to the success of these methods for the finding of satisfactory solutions. The most well known methods have been proposed by Zionts and Wallenius (1976), Hwang et al. (1979), Wierzbicki (1980), Zeleny (1982), Goicoechea et al. (1982), Steuer (1986), Teghem et al. (1986), and Slowinski R. and J. Teghem (1990).

As already mentioned, the multiobjective consideration of the linear programming emerges from the desire of the analyst to render his model more realistic, to take into account more than one optimization criteria. This new form of modeling presupposes the linearity of all the criteria, that is the extra arithmetic coefficients $c_{ij}, i = 1,2,\ldots,n$, $j = 1,2,\ldots,l$ of the objective functions should be found.

It should also be mentioned that a multiobjective decision problem belongs to the category of ill-structured problems. It is a problem in which the optimal solution does not initially exist, but satisfactory solutions can be found through iterative and interactive procedures.

The modern methods of multiobjective linear programming are interactive methods, contain stages of calculation and stages of dialogue between man and machine, that aim at the comprehension of the decision maker of his own preferences and at the determination of the most satisfactory solution.

3.1 Simple methods of multiobjective linear programming

3.1.1 Lexicographic optimization

The method aims at the following algorithmic stages:

1. The decision maker classifies the objectives from the most to the least important: $g_1, g_2, \ldots, g_n$.
2. The objective g_i is maximized over the set F of feasible solutions and let $F^i \subseteq F$ the set of optimal solutions corresponding to the maximization of g_i.
3. Let $F^i \subseteq F^{i-1} \subseteq F^{i-2} \subseteq \ldots \subseteq F^1 \subseteq F$ be the set of optimal solutions of the ith objective maximized over the set F^{i-1} $(\max g_i(x), x \in F^{i-1}), i = 1,2,\ldots,n$
4. If $|F^i| = 1$ or $i=n$ there is only one solution, otherwise, it is set $i=i+1$ and go to stage 2.

Based on this algorithmic presentation it is obvious that the solution of a multiobjective linear programming is completed when a unique solution is found by substitution when all the criteria are examined. It is obvious that in this optimization procedure the classification of the goals based on their importance is significant, while it should be mentioned that the independent examination of each criterion from the others degrades partially the multiobjective character of the problem.

3.1.2 Global criterion method

The global criterion is defined as the method that aims at the composition of n objective functions to one, converting the multiobjective linear problem to a simple optimization problem. This new function is a value function:

$$u(x) = u[g_1(x), g_2(x), ..., g_n(x)] \tag{3.3}$$

Besides the technical difficulties that may arise for the new problem:

$$[\max] u(x) \tag{3.4}$$

subject to $x \in F$

given that the value function can be non-linear, it arises a subject of appropriateness for the specific function as an overall preference model. The last problem of determining the value function of the decision maker leads to the multi-attribute utility theory. In the most simple case the function is considered to be linear expressed as a weighted average of the goals $g_1, g_2, \ldots, g_n$:

$$u(x) = \sum_{i=1}^{n} p_i g_i(x) \tag{3.5}$$

where $p_1, p_2, \ldots, p_n$ are the positive weight coefficients of the goals.

As an overall preference model of a decision maker the above function owes to respect the following preference relations:

$$u(x) > u(y) \Leftrightarrow x \succ y \quad \text{the solution } x \text{ is preferred to } y \tag{3.6}$$

$$u(x) = u(y) \Leftrightarrow x \sim y \quad \text{the solution } x \text{ is indifferent to } y \tag{3.7}$$

It becomes obvious that the use of such a value function for the solution of a multiobjective linear programming depends on the determination of the weight coefficients p_i. For the comprehension of the importance of these coefficients, let two solutions x and y, whose performance to the predetermined goals $g_1, g_2, \ldots, g_n$ differ only in two goals, g_1 and g_i as follows:

$$x : g_1 \quad g_2 \ldots g_i \ldots\ldots g_n \tag{3.8}$$

$$y : g_1\text{-}\Delta \quad g_2 \ldots g_i+1 \ldots g_n \tag{3.9}$$

Based on the performance of the two solutions to the predetermined goals and taking into consideration that both solutions are satisfactory it is concluded that the decision maker is ready to resign Δ units of the goal g_1 in order to win a unit of the goal g_i. Thus, since the two solutions are equivalent, it should be:

$$u(x) = u(y) \Leftrightarrow$$

$$\begin{aligned} &p_1 g_1 + p_2 g_2 + \ldots p_i g_i + \ldots + p_n g_n = \\ &p_1(g_1 - \Delta) + p_2 g_2 + \ldots + p_i(g_i + 1) + \ldots + p_n g_n \end{aligned} \tag{3.10}$$

and thus:

$$\Delta = \frac{p_i}{p_1}, \quad \forall i \tag{3.11}$$

3.1.3 Interactive procedures

In case any multiobjective mathematical programming solution methodology accommodates the need for searching the whole efficient set without referring to the lexicographic optimization and global criterion method, interactive and iterative procedures are being performed. In the first stage of such procedures an initial efficient solution is obtained and it is presented to the decision maker. If this solution is considered acceptable by the decision maker, then the solution procedure stops. If this is not the case, then the decision maker is asked to provide information regarding his preferences on the pre-specified objectives. This information involves the objectives that need to be improved as well as the trade-offs that he is willing to undertake to achieve these improvements. The objective of defining such information is to specify a new search direction for the development of a new improved solution. This process is repeated until a solution is obtained that is in accordance with the decision maker's preferences or until no further improvement of the current solution is possible.

In the international literature several methodologies have been proposed that operate within the above general framework for addressing multiobjective mathematical programming problems. Some well-known examples are the methods developed by Benayoun et al. (1971), Zionts and Wallenius (1976), Wierzbicki (1980), Steuer and Choo (1983), Korhonen (1988), Korhonen and Wallenius (1988), Siskos and Despotis (1989), Lofti et al. (1992). An important part of several methods of multiobjective mathematical programming constitutes the development of the utility function that rules the policy that follows the decision maker, which is then maximized over the set of feasible solutions in order to find the most appropriate efficient solution. The methodology proposed by Siskos and Despotis (1989) is based on this approach and the ADELAIS (Aide à la DEcision pour systèmes Liné-

approach and the ADELAIS (Aide à la DEcision pour systèmes Linéaires multicritères par AIde à la Structuration des préférences) system is implemented.

3.1.4 Goal programming

The goal programming, which constitutes the subject of study of the present volume, deals with performance problems of pre-specified objectives and goals in contrast to multiobjective linear programming that deals with problems of minimization or maximization of various objective functions.

The basic idea of goal programming has been studied by Charnes, Cooper and Ferguson (1955) on executive compensation, while Spronk (1981) provides an extensive discussion of goal programming as well as its applications in the field of financial planning.

Although Charnes, Cooper and Ferguson did not use the term goal programming, they presented a constrained regression idea that embodies the deviation minimizing approach inherent in goal programming. According to Romero (1991), it was not until Charnes and Cooper's 1961 linear programming textbook that the term goal programming appeared. It was not presented as a unique or revolutionary methodology, but as an extension of linear programming. In the Charnes and Cooper book, goal programming was suggested for use in solving unsolvable linear programming problems.

The concept of goal in goal programming is different from that of objective. The meaning of the goal, which constitutes the core of this alternative, differs from the meaning of the objective function, which constitutes the basis of the multicriteria mathematical programming: each objective function indicates simply the direction towards which the existence of satisfactory solutions should be investigated (such as the minimization of cost, maximization of profit, etc.). On the other hand, the clear delimitation of goals allows the evaluation of degree to which each solution corresponds (Keeney and Raiffa, 1993). Contrary to the multicriteria mathematical programming, goal programming techniques do not aim at the immediate optimization of each objective function, but to the search for solutions, which optimize a function of deviations.

3.1.4.1 Goal Programming as an extension of linear programming

Goal programming constitutes of a modification and extension of linear programming. These two programming techniques are similar to the fact that they both represent optimal solutions to goals and constraints. Nevertheless, goal programming and linear programming perform significant differences by giving advantage to goal programming, which is due to the greater scale of problems that is applied (Zeleny, 1982).

Since the origin of goal programming can be traced to linear programming, the foundation of the goal programming model could be based on the assumptions and modeling notation of the linear programming.

The following version of the linear programming model is called the canonical form:

$$Minimize: Z = \sum_{j=1}^{n} c_j x_j \tag{3.12}$$

$$\text{subject to: } \sum_{j=1}^{n} a_{ij} x_j \geq b_i, \text{ for } i = 1,...m \tag{3.13}$$

$$x_j \geq 0, \text{ for } j = 1,...,n \tag{3.14}$$

where $x_1, x_2, ..., x_n$ are nonnegative decision variables and $c_1, c_2, ..., c_n$ are contribution coefficients that represent the marginal contribution to Z for each unit of their respective decision variable.

$a_{ij}, i = 1,...,m, j = 1,...n$ are technological coefficients of the decision variables x_j.

The linear programming model requires the following assumptions (Fang and Puthenpura, 1993):

- Proportionality assumption: Each unit of each decision variable x_j contributes to c_j units of the objective function and a_{ij} units in the ith constraint.
- Additive assumption: The contribution to the objective function and the technological coefficients in the constraints are independent of the values of the decision variables.
- Divisibility assumption: Decision variables are permitted to be noninteger or have fractional values.
- Certainty assumption: All parameters, a_{ij}, b_i and c_j must be known with certainty.

Regardless of the types of constraints included in the linear programming model, the requirements represented by the constraints must be satisfied in order to have a feasible solution. It should also be taken into account that optimization over a feasible solution set of the x_j will satisfy all the constraints in the model. When there are one or more contradictory constraints in a linear programming model, we have an infeasible problem (no solution can be found that satisfies the existing contradictory constraints).

Charnes and Cooper (1961) suggested that each constraint of the linear programming model is a separate function, called a functional. These functionals are viewed as individual objectives or goals to be attained. More precisely, b_i is a set of objectives or goals that should be satisfied in order to have a feasible solution. If b_i is subtracted from both sides of an equality constraint, the functional can be expressed as the absolute value of a linear programming constraint, as follows:

$$f_i(x) = \left| \sum_{j=1}^{n} a_{ij} x_j - b_i \right| \; for\ i = 1,...,m \tag{3.15}$$

Charnes and Cooper (1961) referring to these functionals as goals, suggested that goal attainment is achieved by minimizing their absolute deviation. In these linear programming problems, where deviation in functionals are inevitable, the best solution occurs by minimizing the deviation. Thus, it is possible to obtain a kind of solution where constraints are in conflict with one another.

Recognizing that deviation will exist in unsolvable linear programming problems like an infeasible linear programming problem, Charnes and Cooper (1961) illustrated how that deviation could be minimized by placing the variables representing deviation directly in the objective function of the model. This allows multiple goals to be expressed in a model that will permit a solution to be found. Multiple and conflicting goals are a distinguishing characteristic to describe how a goal programming model differs from a linear programming model. Charnes and Cooper (1977) presented a generally accepted statement of a goal programming model, as follows:

$$Minimize: Z = \sum_{i \in m} (d_i^+ + d_i^-) \tag{3.16}$$

$$\text{subject to: } \sum_{j=1}^{n} a_{ij} x_j - d_i^+ + d_i^- = b_i, \text{ for } i = 1,...,m \tag{3.17}$$

$$d_i^+, d_i^-, x_j \geq 0, \; for\ i = 1,...,m \; j = 1,...,n \tag{3.18}$$

where

d_i^+ is called a positive deviation variable or over-achievement of goal b_i

d_i^- is called a negative deviation variable or under-achievement of goal b_i

b_i is the arithmetic value of goal i

The value of Z is the sum of all deviations. The deviation variables are related to the functionals where:

$$d_i^+ = 1/2\left[\left|\sum_{j=1}^{n} a_{ij}x_j - b_i\right| + (\sum_{j=1}^{n} a_{ij}x_j - b_i)\right] \quad (3.19)$$

$$d_i^- = 1/2\left[\left|\sum_{j=1}^{n} a_{ij}x_j - b_i\right| - (\sum_{j=1}^{n} a_{ij}x_j - b_i)\right] \quad (3.20)$$

More specifically, in goal programming, for each criterion, the decision maker should define the goal that he intends to accomplish. The values-goals that are defined are represented as $s_1, s_2, ..., s_n$ and the model converts all the objective functions to constraints with the input of deviation variables from the goals. The general form of a goal programming model is the following:

$$\min z = \sum_{i=1}^{n} p_i f_i(d_1^-, d_1^+, d_2^-, d_2^+, ..., d_n^-, d_n^+) \quad (3.21)$$

subject to:

$$\sum_{j=1}^{l} c_{ij}x_j + d_i^- - d_i^+ = s_i, \quad i = 1,2,...,n \quad (3.22)$$

$$x \in F \quad (3.23)$$

$$d_i^- \geq 0, \quad d_i^+ \geq 0, \quad i = 1,2,...n \quad (3.24)$$

where, besides the usual data, the follows are also entered:

s_i : the arithmetic value of goal i

p_i : the priority weight of goal i

d_i^+ : over achievement of goal s_i

d_i^- : under achievement of goal s_i

f_i : the linear function of the variables d_i^+ and d_i^-

F : the set of the feasible solutions.

It is obvious that in the goal programming the decision maker determines his goals through an objective function, formulating them on the basis of the following three variables-factors:

- Deviational variables
- Preemptive priority factors
- Weighting of deviational variables

- *Deviational variables*

Contrary to linear programming, which maximizes or minimizes an objective function, goal programming minimizes the deviations from the pre-specified goals, which are defined over the multiple objective functions of the problem. The deviation variables could be represented as d^{+} or d^{-}, corresponding to the positive or negative deviations from the goals.

- *Preemptive priority factors*

For the optimization of the goals with priority levels based on the significance that they have, a classification method of the goals is included into the model. This classification is succeeded through the determination of several priority levels for the deviation variables that correspond to each goal. When the priority level of a goal is equal to one, this means that the corresponding goal is first in the hierarchy and thus it should be accomplished first before the examination of the other goals. The goals with lower priority levels are accomplished first and after the performance of these goals the goals with the largest priority levels are taken into account. That is, the value of the priority level represents the hierarchy of the goals in proportion to their significance.

When Ijiri (1965) had introduced the idea of combining preemptive priorities and weighting in accounting problems, Charnes and Cooper (1977) suggested the goal programming model as:

$$Minimize\ Z = \sum_{i \in m} P_i \sum_{k=1}^{n_i} (w_{ik}^{+} d_i^{+} + w_{ik}^{-} d_i^{-}) \quad (3.25)$$

$$\text{subject to: } \sum_{j=1}^{n} a_{ij} x_j - d_i^{+} + d_i^{-} = b_i, \text{ for } i = 1,..,m \quad (3.26)$$

$$d_i^{+}, d_i^{-}, x_j \geq 0, \text{ for } i = 1,...,m, j = 1,..,n \quad (3.27)$$

where

P_i are the preemptive priority factors that serve only as a ranking symbol meaning that no substitutions across categories of goals will be permitted. It is assumed that the ordering of deviation variables in an objective function, will be minimized in order, where $P_i \geq P_{i+1} \geq \ldots$. It is also assumed that no combination of relative weighting attached to the deviation variables can produce a substitution across categories in the process of choosing the x_j

$w_{ik}^+, w_{ik}^- \geq 0$ represent the relative weights to be assigned to each of the $k=1,\ldots n_i$ different classes within the *i*th category to which the value of P_i is assigned.

- *Weighted deviation variables*

In some cases, it is essential to weigh the deviation variables, which have similar priority level. The result of using the weighted deviation variables is the representation of the relevant significance of the deviation variables with similar priority factors (Lee and Chesser, 1980).

Charnes and Cooper (1977) stated the weighted goal programming model as:

$$Minimize: Z = \sum_{i \in m} (w_i^+ d_i^+ + w_i^- d_i^-) \quad (3.28)$$

$$\text{subject to: } \sum_{j=1}^{n} a_{ij} x_j - d_i^+ + d_i^- = b_i, \text{ for } i = 1,\ldots,m \quad (3.29)$$

$$d_i^+, d_i^-, x_j \geq 0, \text{ for } i = 1,\ldots,m,\ j = 1,\ldots,n \quad (3.30)$$

where

w_i^+ and w_i^- are nonnegative constants representing the relative weight to be assigned to the respective positive and negative deviation variables. The relative weights may be any real number, where the greater the weight the greater the assigned importance to minimize the respective deviation variable to which the relative weight is attached. This model is a nonpreemptive model that seeks to minimize the total weighted deviation from all goals stated in the model.

The proposed methodology takes into account this assignment of relative weights to the deviations for the satisfaction of the goals and constraints proportionally to the demands of the banking managers. Two alternatives are then proposed. According to the first alternative first priority level is given to the solvency goal, second priority level is given to the liquidity goal and third priority level is given to the rest goals including the goals of deposits and loans. In the second alternative first priority level is given to the liquidity goal, second priority level is given to the solvency goal and third priority level to the rest goals.

Concluding, goal programming constitutes an effort of optimization of the set of deviations from pre-specified multiple goals, which are taken into account sequentially and are weighted according to their relevant significance (Zeleny, 1982).

The goal programming methodology could be expanded to the sensitivity analysis and the parameterization of the weighted coefficients of the criteria p_i and the goals b_i.

3.1.5 The optimization role

The scientific preparation of the decisions, a role that has been assigned to the operations research since years, can not be limited to the development of a solution, which is characterized to be optimal according to a mathematical criterion. This could substitute the decision maker with a computer, which could generate optimal solutions.

In order to understand the role of the multiobjective linear programming models, the following should be noted:

- The decision models are based on partly realistic assumptions, which could be overbalanced or permitted to modifications, which were not foreseen during the modeling process.
- An optimal solution is the outcome of a given modeling and could not be identified with the fuzzy and misleading meaning of the optimal decision. Such a solution could not support the real decision, without the recourse to alternative techniques and methods, which are activated after the optimization process (post-optimality analyses).
- The environment, in which the decisions are taken, is characterized by various behaviors so that the limits of the feasible and pursued outcome are not clear.

Bearing these issues in mind, it is clear that the process of decision support begins after the optimization. The most important post optimality techniques are the following.

- Sensitivity analysis, which involves the determination of a range of values for the problem data in which the optimal solution remains unchanged.
- Parametric programming, which involves the characterization of the optimal solution in terms of changes in the problem data.
- Near optimality analysis, which involves the identification of near optimal or alternative optimal solutions.

The most important disadvantage of the first two techniques is that the analysis is marginal, concerns only one coefficient each time assuming that all the other coefficients remain stable, which is not realistic.

One of the most important problems, to which the sensitivity analysis responds, is the calculation of the intervals, into which the coefficients *c*, *b* and *A* of the linear programming could change without the modification of its optimal solution.

3.1.6 Dominance analysis

Referring to the above goal programming, a solution x is dominated by an other solution y if and only if

$$g_i(x) \geq g_i(y) \quad \forall i \tag{3.31}$$

and for at least one criterion i^*:

$$g_{i^*}(x) \succ g_{i^*}(y) \tag{3.32}$$

Generally, a feasible solution $x \in F$, where *F* is the feasible set, is called efficient for the criteria family $\{g_1, g_2, \ldots, g_n\}$ if and only if there is no other feasible solution $y \in F$, which dominates over x.

Based on this formulation, we select, in this study, the non-dominant solutions from the obtained solutions x_i of the goal programming problem, that is the efficient solutions xi for which there are no solutions yj, which dominate over *xi*.

3.1.7 Issues related to goal programming model formulation

Numerous goal programming publications have dealt with difficulties and criticisms on the use of goal programming models. Some of the more notable comments are in Harrald, Leotta, Wallace and Wendell (1978), Hannan (1980), Zeleny (1982), Alvord (1983), Rosenthal (1983), Hannan (1985), Ignizio (1985), Gass (1987), Romero (1991) and Min and Storbeck (1991). Most of these studies involve the issues of dominance, inferiority, efficiency in goal programming solutions, the issues of incommensurability and the use of naïve relative weighting in goal programming models, redundancy and others.

3.1.7.1 Dominance, inferiority and efficiency in goal programming solutions

Goal programming models, similarly to linear programming, often have multiple solutions. Unlike linear programming, the goal programming model can permit a variety of alternative solutions that may allow at least one of the model's goals to be improved without worsening or degrading the others. Zeleny (1982) and other goal programming researchers suggest that such situation represents a major defect in goal programming modeling.

A dominated solution occurs in goal programming, if and only if, an alternative feasible solution can be found that will not reduce deviation in an objective function while reducing deviation of some other goal. Cohon (1978) referred to a nondominated solution as one where no other feasible solution existed that would improve one goal without degrading other conflicting goals. A dominated solution can be called an inferior solution because other superior solutions that yield satisfactory responses exist. Likewise a nondominated solution can be called a noninferior solution because it represents the best solution and not one that is inferior to any other.

The issue of efficiency in goal programming solutions is closely related to dominance and inferiority. According to Pareto (1896), efficiency is at an optimal level if the economic situation of a group of people can not be improved without worsening the economic situation of any one person who makes up the group. This type of optimality is called Pareto efficiency (Romero, 1991). A goal programming solution is said to be Pareto efficient if no other feasible solution can achieve the same or better solution for the group of goals that exist in the objective function, while also being better for one or more other individual objectives that exist in the model. Thus, a goal programming efficient solution must be nondominated solution and a noninferior solution. A goal programming inefficient solution is a dominated solution.

One issue in goal programming model formulation that is directly related to the issues of dominance, inferiority and efficiency concerns the use of linear programming constraints in goal programming models to restrict decision variable values. It is certainly a fact that goal programming permits the inclusion of linear programming constraints into the formulation of any goal programming model. Moore, Taylor, Clayton and Lee (1978) referred to the linear programming constraints included in goal programming models as system constraints. Ignizio (1985) referred to these constraints as a set of rigid constraints.

3.1.7.2 Naïve relative weighting, incommensurability, naïve prioritization and redundancy in goal programming model formulation

The four goal programming model formulation issues of naïve relative weighting, incommensurability, naïve prioritization and redundancy are often related to one another. In fact, any one of these issues can cause formulation problems that in turn often causes the other issues to bring up criticisms against goal programming models.

In any of the weighted goal programming models, the relative weights establish the importance of goals to which they are attached. If the weights do not accurately reflect the true proportioned weight that exists in the decision environment that is being modeled, then we have a situation of naïve relative weighting. Since relative weights are often viewed as a type of utility function, Rosenthal (1983) has argued that weights will almost never reflect the true economic environment they are trying to describe.

The negative impact of naïve relative weighting can be minimized by putting more effort into their calculation. The use of such weighting methods as the analytic hierarchy process (Saaty, 1980), conjoint analysis (Green and Srinivasan, 1990), and even multiple regression analysis can help to improve the accuracy of weighting.

A related issue to relative weighting, though not in a cause-and-effect relationship, is that of incommensurability of goal constraints. In a weighted goal programming model goal constraints are often used to model very different types of goals.

There are many different methodologies and algorithms used to generate solutions for goal programming models. We categorize them into four groups of linear goal programming, integer goal programming, nonlinear goal programming and an other group for all methodologies that do not fit into the other three groups.

Linear Goal Programming

The first linear goal programming algorithm is actually a linear programming algorithm. The methodological proofs for solving linear programming models structured as goal programming problems can be found in Charnes and Cooper (1961). With the improvements of preemption, the generalized inverse approach and the use of simplex based algorithm by Ijiri (1965), as well as the publication of a software program by Lee (1972), increased linear goal programming research in methodological improvements. Ignizio (1976, 1982), Ijiri (1965), Lee (1972) and Schniederjans (1984) presented the basic algorithms used to solve the weighted goal programming and preemptive goal programming.

Integer Linear Goal Programming

In goal programming problem situations where decision variables are restricted to integer values, special integer goal programming methodologies were developed. Most of the goal programming methodologies are based on integer linear programming methodologies. In most mixed integer linear programming problems one of the most common integer methodologies is the branch-and-bound solution method. Arthur and Ravindran (1980) developed their branch-and-bound integer goal programming algorithm on the same linear programming algorithm.

Nonlinear goal programming

According to Saber and Ravindran (1993) there are four major approaches to nonlinear goal programming: (a) simplex based nonlinear goal programming; (b) direct search based nonlinear goal programming; (c) gradient search based nonlinear goal programming; (d) interactive approaches to nonlinear goal programming.

Simplex based nonlinear goal programming approaches include the method of approximation programming, developed by Griffith and Steward (1961) adapted by Ignizio (1976) for goal programming. This methodology permits nonlinear goal constraints to be included in a goal programming model.

Another simplex based approach to nonlinear goal programming is separable programming. This approach was originally developed by Miller

(1963) and allows nonlinear goal constraints to be included in the goal programming model by restricting the range of the decision variables into separable functions that are assumed linear. This methodology is based on piecewise linear approximations.

Another simplex based approach to nonlinear goal programming is quadratic goal programming, presented firstly by Beale (1967). Quadratic goal programming permits quadratic goal constraints and quadratic deviation variables in the objective function.

Direct search based nonlinear goal programming methods use some type of logical search pattern or methods to obtain a solution that may or may not be the best satisfactory solution. The logic process is based on repeated attempts to improve a given solution by evaluating its objective function and goal constraints. The basic search idea originated with Box (1965), but was applied to goal programming by many others including Nanda, Kothari and Lingamurthy (1988).

Gradient based nonlinear goal programming methods use calculus or partial derivatives of the nonlinear goal constraints or the objective function to determine the direction in which the algorithm seeks a solution and the amount of movement necessary to achieve that solution. While gradient based methods are generally more efficient in obtaining a solution, they may not be appropriate for goal programming models whose goal constraints or objective function is nondifferentiable.

Based on Zoutendijk's (1960) algorithm Lee (1985a), Newton (1985) and Olson and Swenseth (1987) developed a version of the gradient method for goal programming called the chance constraint method. The chance constraint method allows parameters to be distributed along a probability distribution. The introduction of the probability distribution is where this methodology obtained its probabilistic or chance name. The use of the chance constraint method requires the assumption that the technological coefficients are normally distributed.

Another gradient based method is the partitioning gradient method. Developed for linear goal programming by Arthur and Ravindran (1978) using a simplex based approach, this methodology can be highly efficient in obtaining nonlinear goal programming solutions. It works on the basis of finding smaller problems that lead to an optimal solution. By solving these smaller problems and eliminating decision variables from the model, the size of the model is reduced. A special version of the gradient based method is called the decomposition method, which can solve linear goal programming or nonlinear goal programming problems. It is based on a version of Dantzig and Wolfe (1960) linear programming decomposition method, where large models are decomposed into smaller models whose solution will be used to generate the solution to the original larger model. Ruefi (1971), Sweeney, Winkafsky, Roy and Baker (1978), Lee (1983) and Lee and Rho (1979a, 1979b, 1985, 1986) present algorithms and research on the decomposition method for goal programming.

A special type of nonlinear goal programming methodology is stochastic goal programming. In a stochastic goal programming model there are probability distributions that describe either the model parameters or the model's structure.

Interactive approaches to nonlinear goal programming or interactive goal programming can be defined as a collection of methodologies that are based on progressive articulation of a decision maker's preferences in a decision environment (Dyer, 1972).The decision maker using interactive goal programming will be led to a better solution by interactively comparing a given solution. This makes interactive goal programming a sequential search process, that involves periodic feedback to the decision maker to guide the direction of the search. The term sequential goal programming is often used with the interactive approach to better describe the step-wise nature of this methodology.

3.1.7.3 Other goal programming algorithms and methodology

Interval goal programming, fractional goal programming, duality solution and fuzzy goal programming are four other algorithms based methodologies represented extensively in goal programming literature. Each of these methodologies are often used with linear goal programming, integer goal programming and nonlinear goal programming models.

Interval goal programming

Interval goal programming allows parameters, particularly the right-hand side goal values to be expressed on an interval basis. This method is based on interval linear programming, where an upper boundary b_u and lower boundary b_l are stated as follows:

$$b_l \leq a_{ij} x_j \leq b_u \tag{3.33}$$

Thus, the interval goal programming can be accomplished with two goal constraints:

$$a_{ij}x_j - d_u^+ + d_u^- = b_u \tag{3.34}$$

$$a_{ij}x_j - d_l^+ + d_l^- = b_l \tag{3.35}$$

where the d_u^+ and d_l^- are both minimized in the objective function and the other deviation variables are free to permit some compromised values for the resulting right-hand-side value.

Fractional goal programming

Fractional goal programming is a methodology used when modeling ratios. In cases of modeling return on investment problems, market share problems or percentage type problems, fractional goal programming could be the most appropriate of the goal programming methodologies. Fractional goal programming is also an extension of linear programming, called fractional linear programming (Marto, 1964; Bitran and Novaes, 1973).

Duality solution

It has been shown that goal programming models can be solved more efficiently and without some computational problems by solving the dual formulation of a goal programming model (Dauer and Krueger, 1977; Ignizio, 1985). El-Dash and Mohamed (1992) present an interesting extension of this method to sequential nonlinear goal programming.

Fuzzy goal programming

Fuzzy goal programming is based on fuzzy set theory, used to describe imprecise goals. These goals are usually associated with objective functions and reflect both a weighting and range of goal achievement possibilities.

As it is ascertained from the above sections, the specific simplicity that governs the use of the goal programming techniques has rendered them popular among operation researchers for the solution of many practical optimization problems subject to multiple goals and constraints. An historical review to the goal programming theory and its applications to the various fields of scientific research is presented by Zanakis and Gupta (1985), Romero (1986, 1991), Schniederjans (1995) and Aouni and Kettani (2001).

4. INTEREST RATE SIMULATION ANALYSIS

The transition to the new monetary and economic environment, which is created, requires the development and application of high quality systems for the evaluation of the risks faced by financial institutions. More specifically the banks should manage the interest rate risk, the operating risk, the credit risk, the market risk, the foreign exchange risk, the liquidity risk and the country risk. A series of norms is applied to each of these categories for their better confrontation (Chapter 1).

As it was already mentioned (Chapter 1), the interest rate risk indicates the effect of the changes to the net profit margin between the deposit and borrowing values, which are evolved as a consequence of the deviations to the dominant interest rates of assets and liabilities. When the interest rates diminish, the banks accomplish high profits since they can refresh their liabilities to lower borrowing values. The reverse stands to high borrowing values. It is obvious, that the changes of the inflation have a relevant impact on the above sorts of risk.

Interest rate risk contributes significantly to the composition of a bank's assets and liabilities. Interest rate risk categories constitute the deposits, loans and bonds rates. The interest rates of all the loan categories (as well as the deposit categories) are adjusted by the monetary authorities contributing to the abetting of the economic development. The basic goal of the policy determination of the loan interest rates is the reinforcement of the demand for credits from several branches or for subsidization with the institution of low loan interest rates.

The measurement techniques of the changes in interest rates include gap analysis, duration analysis, as well as simulation models. It is mentioned (Chapter 1) that the most widely used technique for the financial risk management and more specifically for the interest rate risk management is the Monte Carlo simulation, which is used in the present study for the creation of scenarios to the interest rate values in order to take into account the interest rate risk, which is described in the next section.

4.1 Monte Carlo simulation

Considering, as it was already mentioned, the interest rate risk as the basic uncertainty parameter to the determination of a bank asset liability management strategy, the crucial question that arises concerns the determination of the way through which this factor of uncertainty affects the profitability of the pre-specified strategy. The estimation of the expected return of the pre-specified strategy and of its variance can render a satisfactory response to the above question.

The use of Monte Carlo techniques constitutes a particular widespread approach for the estimation of the above information (expected return – variance of bank asset liability management strategies). Monte Carlo simulation consists in the development of various random scenarios for the uncertain variable (interest rates) and the estimation of the essential statistical measures (expected return and variance), which describe the effect of the interest rate risk to the selected strategy. The general procedure of implementation of Monte Carlo simulation based on the above is presented in Figure 3.2.

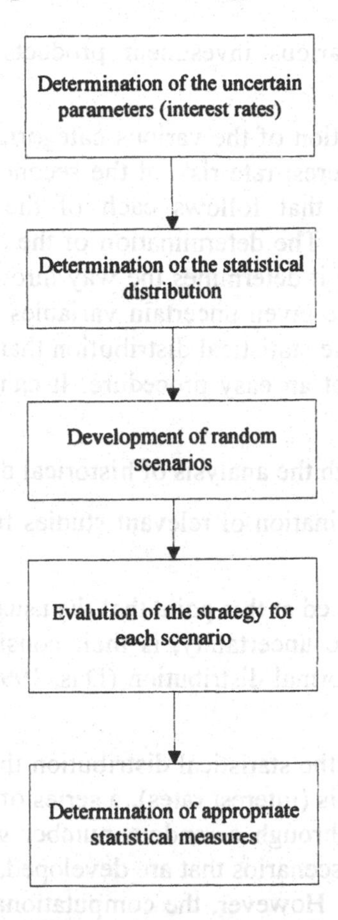

Figure 3.2: General Monte Carlo simulation procedure for the evaluation of the asset liability management strategies

During the first stage of the procedure the various categories of the interest rate risks are identified. The risk and the return of the various data of bank asset and liability are determined from the different forms of interest rates. For example, the investments of a bank to government or corporate bonds are determined from the interest rates that prevail in the bond market, which are affected so by the general economic environment as by the rules of demand and supply. Similarly, the deposits and loans of the bank are determined from the corresponding interest rates of deposits and loans, which are assigned by the bank according to the conditions that prevail to the bank market. At this stage, the categories of the interest rates, which constitute crucial uncertain variables for the analysis, are detected. The determined interest rates categories depend on the type of the bank. For example, for a decisive commercial bank, the deposit and loan interest rates have a role, whereas for an investment bank more emphasis is given to the interest rates and the returns of various investment products (repos, bonds,

and the returns of various investment products (repos, bonds, interest-bearing notes, etc.).

After the determination of the various categories of interest rates, which determine the total interest rate risk, at the second stage of the analysis the statistical distribution that follows each of the pre-specified categories should be determined. The determination of the statistical distribution is a significant issue, since it determines the way through which a series of random scenarios over the given uncertain variables is developed. Generally, the determination of the statistical distribution that follows each category of the interest rates is not an easy procedure. It can be accomplished in two ways:

1. Empirically through the analysis of historical data.
2. Through the examination of relevant studies from the international bibliography.

It should be mentioned at this point that the usual approach for the modeling of the interest rate uncertainty, is their consideration as random variables, which follow normal distribution (Das, 1998; Smithson, 1998; Ross, 1999).

Having determined the statistical distribution that describes the uncertain variables of the analysis (interest rates), a series of random independent scenarios is developed, through a random number generator. Generally, the largest the number of scenarios that are developed, the more reliable conclusions can be derived. However, the computational effort increases significantly, since for each scenario the optimal asset liability strategy should be determined and moreover its evaluation for each other scenario should take place. Thus, the determination of the number volume N of simulations (scenarios), which will take place should be determined, taking into account both the reliability of the results and the available computational resources.

For each scenario s_i ($i = 1, 2, \ldots, N$) over the interest rates the optimal asset liability management strategy Y_i is determined through the solution of the goal programming problem. It is obvious that this strategy is not expected to be optimal for each of the other scenarios s_j ($j \neq i$). Therefore the results obtained from the implementation of the strategy Y_i under the rest $N-1$ possible scenarios s_j should be evaluated. The evaluation of the results can be implemented from various directions. The most usual is the one that uses the return. Representing as r_{ij} the outcome (return) of the strategy Y_i under the scenario s_j, the expected return $\overline{r}_i$ of the strategy can be easily determined based on all the other $N-1$ scenarios s_j ($j \neq i$), as follows:

$$\overline{r}_i = \frac{1}{N-1} \sum_{j=1, j \neq i}^{N} r_{ij} \tag{3.36}$$

At the same time, the variance σ_i^2 of the expected return can be determined as a risk measure of the strategy Y_i, as follows:

$$\sigma_i^2 = \frac{1}{N-1} \sum_{j=1, j \neq i}^{N} \left(r_{ij} - \overline{r}_i \right)^2 \tag{3.37}$$

These two statistical measures (average and variance) contribute to the extraction of useful conclusions concerning the expected efficiency of the asset liability management strategy, as well as the risks that it carries. Moreover, these two basic statistical measures can be used for the expansion of the analysis of the determination of other useful statistical information, such as the determination of the confidence interval for the expected return, the quantiles, etc.

Chapter 4

Application

1. DESCRIPTION OF THE SAMPLE DATA

The present volume, which uses data from a large commercial bank of Greece, presents an asset liability management (ALM) methodology into a stochastic environment of interest rates in order to select the best direction strategies to the banking financial planning.

More precisely, the goal programming model of this study was developed in terms of a one-year time horizon. The model used balance sheet and income statement information for the previous year of the year 20X0 to produce a future course of ALM strategy for the year 20X0. As far as model variables are concerned, we used variables familiar to management and facilitated the specification of the constraints and goals. For example, goals concerning measurements such as liquidity, return and risk have to be expressed in terms of utilized variables.

The variables used in the specification of the model were taken directly from the financial statement of a commercial bank of Greece for the previous year of 20X0. 42 structural variables were used, of which 22 correspond to assets (X_i, i=1,...,22) and 20 to liabilities (Y_j, j=1,...,20), as they are presented in Table 4.1.

Table 4.1: The decision variables of the goal programming formulation

X_1:	Cash	Y_1:	Due to credit institutions
X_2:	Cheques receivable	Y_2:	Due to credit institutions with agreed maturity
X_3:	Deposits in the Bank of Greece	Y_3:	Commitments arising out of sale and repurchase transactions
X_4:	Treasury bills and other securities issued by the Greek State	Y_4:	Deposits repayable on demand
X_5:	Other Treasury bills and securities	Y_5:	Saving deposits
X_6:	Interbank deposits and loans repayable on demand	Y_6:	Deposits with agreed maturity
X_7:	Other interbank deposits and loans	Y_7:	Cheques and orders payable
X_8:	Loans and advances to customers maturing within one year	Y_8:	Commitments arising out of sale and repurchase transactions (customer amounts)
X_9:	Loans and advances to customers maturing after one year	Y_9:	Dividends payable
X_{10}:	Other receivables	Y_{10}:	Income tax and other taxes payable
X_{11}:	Securities issued by the Greek State	Y_{11}:	Withholdings in favor of social security funds and other third parties
X_{12}:	Other securities	Y_{12}:	Other liabilities
X_{13}:	Shares and other variable-yield securities	Y_{13}:	Accruals and deferred income
X_{14}:	Investments in non-affiliates	Y_{14}:	Accrued interest on time deposits
X_{15}:	Investments in affiliates	Y_{15}:	Other accrued expenses of the year
X_{16}:	Other assets	Y_{16}:	Provisions for staff retirement indemnities
X_{17}:	Deferred charges	Y_{17}:	Other provisions for liabilities and charges
X_{18}:	Accrued income state bonds	Y_{18}:	Loans of reduced indemnity
X_{19}:	Accrued income other bonds	Y_{19}:	Share Capital
X_{20}:	Accrued income loans and advances	Y_{20}:	Retained earnings
X_{21}:	Other accrued income		
X_{22}:	Fixed assets		

2. FORMULATION OF THE PROBLEM

Before proceeding to the goal programming formulation and for the better comprehension of the mathematical formulation of the problem it would be appropriate to describe the constraints and goals that were used.

2.1 Constraints

Strategy and policy constraints of the bank

Certain constraints are imposed by the banking regulation on particular categories of accounts. Specific categories of asset accounts (X') and liability accounts (Y') are detected and the minimum and maximum allowed limit for these categories are defined based on the strategy and policy that the bank intends to follow, as described below:

$$\mathrm{K\Phi}_{X'} \leq X' \leq \mathrm{A\Phi}_{X'} \tag{4.1}$$

$$\mathrm{K\Phi}_{Y'} \leq Y' \leq \mathrm{A\Phi}_{Y'} \tag{4.2}$$

where $\mathrm{K\Phi}_{X'}$ $(\mathrm{K\Phi}_{Y'})$ is the low bound of specific asset accounts X' (liability accounts Y')

$\mathrm{A\Phi}_{X'}$ $(\mathrm{A\Phi}_{Y'})$ is the upper bound of specific asset accounts X' (liability accounts Y')

At the present study, taking into account the historical data of the bank, as well as the opinion of the bank managers and realizing an empirical analysis for the 5 previous economic years of the bank, it becomes evident that the average growth rate of loans is 38%. Based on the above, the total loans $(X_8+X_9+X_{10})$ are not expected to maintain at least the previous year's level (7,632,392 €) and cannot rise by more than 38% in relation to these levels, as follows:

$$X_8 + X_9 + X_{10} \geq 7{,}632{,}392 \tag{4.3}$$

$$X_8 + X_9 + X_{10} \leq 1.38 \times 7{,}632{,}392 \tag{4.4}$$

Similarly, based on empirical analysis, we define the following constraints

$$Y_4 + Y_5 + Y_6 + Y_7 + Y_8 \geq 12{,}348{,}981 \tag{4.5}$$

$$Y_4 + Y_5 + Y_6 + Y_7 + Y_8 \leq 1.28 \times 12{,}348{,}981 \tag{4.6}$$

which state that the total deposits are not expected to increase by more than 28%, which is the average growth rate of deposits, above the previous year's levels (12,348,981 €) and cannot be lower than that.

Since the major part of the capital of the commercial banks consists of the share capital the variable Y_{19} is selected for the development of the asset liability management model. The basic factors for the forecasting of the bank capital structure refer to the evaluations regarding to the development of the economic and banking environment and the variables that affect separately each bank. Based on the above and taking into account that the share capital cannot be reduced, the following constraint reflects the prohibition that is imposed to the low bound of the share capital, which value is set larger or equal to 1,052,384, which corresponds to the value of the share capital of the previous year.

$$Y_{19} \geq 1{,}052{,}384 \tag{4.7}$$

Moreover, the ratio retained earnings to total assets (Y_{20}) indicates the profitability of the total assets. Generally, this index indicates how efficient is the bank asset management and how profitable are the investment decisions that are taken into account from the head departments. In the present study, the value of this index, as it is presented at the following constraint, is set larger or equal to 2.27%. This value raised, as it was already mentioned from the historical data of past values for the calculation of the average growth of the ratios retained earnings to total assets.

$$Y_{20} \geq 2.27\% \times \sum_{i=1}^{22} X_i \tag{4.8}$$

Structural constraints

This category of constraints includes those that contribute to the structure of the balance sheet and especially to the performance of the equation Assets=Liabilities + Net Capital.

The bank management should determine specific goals, such as the desirable structure of each financial institution's assets and liabilities for the units of surplus and deficit, balancing the low cost and the high return. The structure of assets and liabilities is significant, since it affects swiftly the income and profits of the bank. It constitutes one of the most important decision takings from the bank managers' part and it is described by the following constraint

$$\sum_{i=1}^{n} X_i = \sum_{j=1}^{m} Y_j \tag{4.9}$$

$\forall i = 1,...,n$, where n is the volume of asset variables

$\forall j = 1,...,m$, where m is the volume of liability variables

X_i : the element i of assets

Y_j : the element j of liabilities

Several constraints that are imposed by the European Central Bank and the Bank of Greece belong to this category of structural constraints and are described as follows.

$$\sum_{j \in \Pi_{Y''}} Y_j - a \sum_{i \in E_{X''}} X_i = 0 \tag{4.10}$$

where

Y_j: the element j of liability $\forall j \in \Pi_{Y''}$, where $\Pi_{Y''}$ are specific categories of liability accounts

X_i : the element i of asset $\forall i \in E_{X''}$, where $E_{X''}$ are specific categories of asset accounts

α : the desirable value of specific asset and liability data

The following constraints

$$Y_4 + Y_5 + Y_6 + Y_7 + Y_8 - 1.99(X_8 + X_9 + X_{10}) = 0 \tag{4.11}$$

$$Y_4 + Y_5 + Y_6 + Y_7 + Y_8 - 2.29(X_4 + X_5 + X_{11} + X_{12} + X_{13}) = 0 \tag{4.12}$$

$$Y_4 + Y_5 + Y_6 + Y_7 + Y_8 - 5.67X_3 = 0 \tag{4.13}$$

are derived from the obligation of the bank to reserve a specific amount of its deposits in a special interest-bearing account at the Bank of Greece (X_3), as well as in interest-bearing government bonds(X_4+X_5+X_{11}+X_{12}+X_{13}). Moreover, a percentage of private deposits is directed towards loans for public sector corporations (X_8+X_9+X_{10}).

The constraint

$$\sum_{i=1}^{22} X_i - \sum_{j=1}^{20} Y_j = 653{,}116 \tag{4.14}$$

defines the equality relationship between assets and liabilities and net worth. The amount 653,116 € refers to the amount of capital that is assumed to be stable. This category of accounts is used for the distribution of dividends in the future, during which the financial institution may not realize profits. Moreover, this account may exist one year and not in the next one. For this reason, the above amount is assumed to be stable and at the previous year's levels.

The following constraint

$$\sum_{i=1}^{22} X_i \leq 1.30 \times 17{,}327{,}046 \tag{4.15}$$

assumes that the total assets are expected to increase by more than 30% above the previous year's levels, as it arises from the average growth rate of the total assets for the past years.

2.2 Goals

Each financial institution sets a specific amount of goals regarding the demands of the bank managers and the policy that it wishes to follow.

The solvency goal, which is described below, is used as a risk measure and is defined as the ratio of the bank's equity capital to its total weighted assets. The weighting of the assets reflects their respective risk, greater weights corresponding to a higher degree of risk. This hierarchy takes place according to the determination of several degrees of significance for the variables of assets and liabilities. That is, the variables with the largest degrees of significance correspond to categories of the balance sheet accounts with the highest risk stages.

$$\sum_{j \in \Pi_1} Y_j - \sum_{i \in E} w_i X_i - d_s^+ + d_s^- = k_1 \quad (4.16)$$

where

Π_1 : the liability set that includes the equity

E : the total assets

X_i : the element i of assets

Y_j : the element j of liability

w_i : the degree of riskness of the assets

k_1 : the solvency ratio defined from the European Central Bank

d_s^+ : the over-achievement of the solvency goal s

d_s^- : the under-achievement of the solvency goal s

As was already mentioned in Chapter 1 a basic policy of the commercial banks is the management of their liquidity and specifically the measurement of their needs that is relative to the progress of deposits and loans. The liquidity goal is defined as the ratio of liquid assets to current liabilities and indicates the liquidity risk, that is indicates the possibility of the bank to respond to its current liabilities with a security margin, which allows the probable reduction of the value of some current data and it is described as follows.

$$\sum_{i \in E_x} X_i - k_2 \sum_{j \in \Pi_k} Y_j + d_l^- - d_l^+ = 0 \quad (4.17)$$

where

X_i : the element i of asset

Y_j : the element j of liability

k_2 : the liquidity ratio defined from the bank policy

E_χ : the set of total assets, which includes the loans

Π_κ : the set of liabilities that includes the deposits

d_l^+ : the over-achievement of the liquidity goal l

d_l^- : the under-achievement of the liquidity goal l

Moreover, the bank aims at the maximization of its efficiency that is the accomplishment of the largest possible profit from the best placement of its funds. Its aim is the maximization of its profitability and therefore precise and consistent decisions should be taken into account during the bank management. These decisions will guarantee the combined effect of all the variables that are included on the calculation of the profits. This decision taking gives emphasis to several selected variables that are related to the bank management, such as to the management of the difference between the asset return and the liability cost, the expenses, the liquidity management and the capital management. The following goal determines that for the total expected return based on the expected returns for all the assets R^X and liabilities R^Y.

$$\sum_{i=1}^{n} R_i^X X_i - \sum_{j=1}^{m} R_j^Y Y_j - d_r^+ + d_r^- = k_3 \tag{4.18}$$

where

X_i : the element i of asset, $\forall i = 1,\ldots,n$

Y_j : the element j of liability, $\forall j = 1,\ldots,m$

R_i^X : the expected return of the element i of asset, $\forall i = 1,\ldots,n$

R_j^Y : the expected return of the element j of liability, $\forall j = 1,\ldots,m$

k_3: the expected value for the goal for the overall return of the assets and liabilities

d_r^+ : the over-achievement of the return goal r

d_r^- : the under-achievement of the return goal r

Beside the goals of solvency, liquidity and return of assets and liabilities, the bank could determine other goals that concern specific categories of assets and liabilities, in proportion to the demands and preferences of the bank managers. These goals, in their general form, are described as follows:

$$\sum_{i \in E_p} X_i + d_p^- - d_p^+ = l_p, \ \forall p \tag{4.19}$$

or

$$\sum_{j\in\Pi_p} Y_j + d_p^- - d_p^+ = l_p, \forall p \tag{4.20}$$

where

p constitutes the goal imposed from the bank

X_i : the element i of asset, $\forall i = 1,...,n$

Y_j : the element j of liability, $\forall j = 1,...,m$

d_p^+ : the over achievement of the goal p

d_p^- : the under achievement of the goal p

l_p : the desirable value goal for the constraint-goal p, which defines the bank

Taking into account all the above, the bank strategy and the particularities of the environment into which the bank activates and based on the data of a large commercial bank, the goals of solvency, liquidity and return of assets and liabilities as well as the goals of deposits and loans and the goals that refer to specific asset variables are modulated and are described as follows.

Solvency goal

According to the proposal of the Commission of the European Communities (1989), this ratio must be greater than or equal to 8% in order to guarantee the required solvency.

$$Y_{19} - 0.3349Y_{20} - 0.2X_4 - 0.5X_8 - 0.7X_9 - 0.5X_{10} - 0.2X_{11} - 0.4X_{12} - X_{13} - d_1^+ + d_1^- = 8\% \tag{4.21}$$

Liquidity goal

According to the bank policy this ratio should be approximately 0.6 and no higher than 0.6, indicating that at least half of the total capital of the bank should be derived from the liquid current data of the bank and not from the deposits in order to avoid the liquidity risk.

$$\sum_{i=1}^{21} X_i - 0.6\sum_{j=1}^{18} Y_j - d_2^+ + d_2^- = 0 \tag{4.22}$$

Deposit goal

The drawing of capital, especially from the deposits constitutes a major part of commercial bank management, as mentioned in Chapter 1. All sorts of deposits constitute the major source of capital for the commercial banks, in order to proceed to the financing of the economy, through the financing of firms. Thus, it is given special significance to the deposits goal.

Taking into account the development and the change of the deposits of the specific bank and generally of the bank branch and the economy in macroeconomic level, the average growth rate of deposits for the last past five year arises approximately to 28%. Thus, at the present study, we assume that the desirable value target of the deposits is at least 28% higher than the previous year's levels, as described below:

$$Y_4 + Y_5 + Y_6 + Y_7 + Y_8 - d_4^+ + d_4^- = 1.28 \times 12{,}348{,}981 \tag{4.23}$$

Moreover, similarly to above, taking into account the historical data of past years of the bank for the relation of deposits to the total assets, we assume that the average growth rate of the ratio deposits to total assets should be at least 73.31%.

$$Y_4 + Y_5 + Y_6 + d_{10}^- - d_{10}^+ = 73.31\% \times 17{,}327{,}046 \tag{4.24}$$

Loan goal

Similarly to the deposits is the desire of the banks to provide loans. Taking into account the historical data of the loans of the previous years of the bank, the desirable value target for the loans granted is set at 38% above the previous year's level, reflecting the management decision to maintain the forecast for deposit growth in the overall economy.

$$X_8 + X_9 + X_{10} - d_3^+ + d_3^- = 1.38 \times 7{,}632{,}392 \tag{4.25}$$

Goal of asset and liability return

This goal defines the goal for the overall expected return of the selected asset-liability strategy over the year of the analysis. This goal is set at 30%, which is the average growth rate and arises from the data of the previous years, and is defined on the basis of the expected returns for all assets R^X and liabilities R^Y.

$$\sum_{i=1}^{22} R_i^X X_i - \sum_{j=1}^{20} R_j^Y Y_j - d_5^+ + d_5^- = 30\% \times 17{,}327{,}046 + 653{,}116 \tag{4.26}$$

Other goals

Finally, it should be noted that the above goal programming model incorporates also the following goals.

$$X_1 - 0.01 \times 17{,}327{,}046 + d_6^- - d_6^+ = 0 \tag{4.27}$$

$$X_2 - 0.004 \times 17{,}327{,}046 + d_7^- - d_7^+ = 0 \tag{4.28}$$

$$X_3 - 0.14 \times 17{,}327{,}046 + d_8^- - d_8^+ = 0 \tag{4.29}$$

$$X_{22} - 0.015 \times 17{,}327{,}046 + d_9^- - d_9^+ = 0 \tag{4.30}$$

reflecting that variables such as cash, cheques receivable, deposits to the Bank of Greece and fixed assets, should remain at the levels of previous years.

More analytically, it is known that the fixed assets are the permanent assets, which have a natural existence, such as buildings, machines, locations and equipment, etc. Intangible assets are the fixed assets, which have no natural existence but constitute rights and benefits. They have significant economic value, which sometimes is larger than the value of the tangible fixed assets. These data have stable character and are used productively by the bank for the regular operation and performance of its objectives. Since the fixed assets, tangible or intangible, are presented at the balance sheet at their book value that is the initial value of cost minus the depreciation till today, it is assumed that their value does not change during the development of the present methodology.

The variables X_1, X_2, X_3 refer to the account “Cash and deposits in the Bank of Greece”. This account involves the deposits of the banks in the Central Bank, which are created for specific purposes, such as the financing of public firms, with the decisions of the monetary authorities. Moreover, it includes the cash, the money transfers, the cheques receivable that are returned from the compensation office. Because these data do not present significant changes from one year to another and do not affect the structure of the balance sheet it is assumed that their price varies at the previous year’s levels.

2.3 Mathematical formulation

As already mentioned, the major advantage of a goal programming technique is its great flexibility, which enables the decision maker the possibility to incorporate easily numerous variations of constraints and goals.

Developing an asset liability management system in the present study, the goal programming formulation, in its general form, can be expressed as follows:

$$\min z = \sum_{P} p_k (d_k^- + d_k^+) \tag{4.31}$$

subject to constraints:

$$\mathrm{K}\Phi_{X'} \leq X' \leq \mathrm{A}\Phi_{X'} \tag{4.32}$$

$$\mathrm{K}\Phi_{Y'} \leq Y' \leq \mathrm{A}\Phi_{Y'} \tag{4.33}$$

$$\sum_{i=1}^{n} X_i = \sum_{j=1}^{m} Y_j \quad \forall i = 1,\ldots,n, \ \forall j = 1,\ldots,m \tag{4.34}$$

$$\sum_{j \in \Pi_{Y''}} Y_j - a \sum_{i \in \mathrm{E}_{X''}} X_i = 0 \tag{4.35}$$

$$\sum_{j \in \Pi_1} Y_j - \sum_{i \in \mathrm{E}} w_i X_i - d_s^+ + d_s^- = k_1 \tag{4.36}$$

$$\sum_{i \in \mathrm{E}_x} X_i - k_2 \sum_{j \in \Pi_k} Y_j + d_l^- - d_l^+ = 0 \tag{4.37}$$

$$\sum_{i=1}^{n} R_i^X X_i - \sum_{j=1}^{m} R_j^Y Y_j - d_r^+ + d_r^- = k_3 \tag{4.38}$$

$$\sum_{i \in \mathrm{E}_p} X_i + d_p^- - d_p^+ = l_p, \ \forall p \tag{4.39}$$

$$\sum_{j \in \Pi_p} Y_j + d_p^- - d_p^+ = l_p, \ \forall p \tag{4.40}$$

$X_i \geq 0$, $Y_j \geq 0$, $d_k^+ \geq 0, d_k^- \geq 0$, for all i=1, …,n ,j=1, …, m, $k \in P$ (4.41)

where

X_i : the element i of asset, $\forall i = 1,\dots,n$, n is the number of asset variables

Y_j : the element j of liability, $\forall j = 1,\dots,m$, m is the number of liability variables

$K\Phi_{X'}$ ($K\Phi_{Y'}$): is the low bound of specific asset accounts X' (liability Y')

$A\Phi_{X'}$ ($A\Phi_{Y'}$): is the upper bound of specific asset accounts X' (liability Y')

$E_{X''}$: specific categories of asset accounts

$\Pi_{Y''}$: specific categories of liability accounts

α : the desirable value of specific asset and liability data

Π_1 : the liability set, which includes the equity

E : the set of assets

w_i : the degree of riskness of the asset data

k_1 : the solvency ratio, as it is defined from the European Central Bank.

k_2 : the liquidity ratio, as it is defined from the bank policy

E_χ : the set of asset data, which includes the loans

Π_κ : the set of liability data, which includes the deposits

R_i^X : the expected return of the asset i , $\forall i = 1,\dots,n$

R_j^Y : the expected return of the liability j, $\forall j = 1,\dots,m$

k_3: the expected value for the goal of asset and liability return

P : the goal imposed from the bank

L_p : the desirable value goal for the goal constraint p defined by the bank

d_k^+ : the over-achievement of the goal k, $\forall k \in P$

d_k^- : the under-achievement of the goal k, $\forall k \in P$

p_k : the priority degree (weight) of the goal k

Taking into account the above formulation of the goal programming, as well as the data of the specific commercial bank of Greece, the proposed goal programming formulation, including the constraints and goals as were described above, can be expressed as follows:

$$\text{Min } z = \sum_{k=3}^{10} d_k^+ + \sum_{k=3}^{10} d_k^- + 2d_2^- + 3d_1^- \tag{4.42}$$

subject to the constraints

$$X_8 + X_9 + X_{10} \geq 7{,}632{,}392 \tag{4.43}$$

$$X_8 + X_9 + X_{10} \leq 1.38 \times 7{,}632{,}392 \tag{4.44}$$

$$Y_4 + Y_5 + Y_6 + Y_7 + Y_8 \geq 12{,}348{,}981 \tag{4.45}$$

$$Y_4 + Y_5 + Y_6 + Y_7 + Y_8 \leq 1.28 \times 12{,}348{,}981 \tag{4.46}$$

$$Y_{19} \geq 1{,}052{,}384 \tag{4.47}$$

$$Y_{20} \geq 2.27\% \times \sum_{i=1}^{22} X_i \tag{4.48}$$

$$Y_4 + Y_5 + Y_6 + Y_7 + Y_8 - 1.99(X_8 + X_9 + X_{10}) = 0 \tag{4.49}$$

$$Y_4 + Y_5 + Y_6 + Y_7 + Y_8 - 2.29(X_4 + X_5 + X_{11} + X_{12} + X_{13}) = 0 \tag{4.50}$$

$$Y_4 + Y_5 + Y_6 + Y_7 + Y_8 - 5.67X_3 = 0 \tag{4.51}$$

$$\sum_{i=1}^{22} X_i - \sum_{j=1}^{20} Y_j = 653{,}116 \tag{4.52}$$

$$\sum_{i=1}^{22} X_i \leq 1.30 \times 17{,}327{,}046 \tag{4.53}$$

$$Y_{19} - 0.3349Y_{20} - 0.2X_4 - 0.5X_8 - 0.7X_9 - 0.5X_{10} - 0.2X_{11} - 0.4X_{12} - X_{13} - d_1^+ + d_1^- = 8\% \tag{4.54}$$

$$\sum_{i=1}^{21} X_i - 0.6\sum_{j=1}^{18} Y_j - d_2^+ + d_2^- = 0 \qquad (4.55)$$

$$X_8 + X_9 + X_{10} - d_3^+ + d_3^- = 1.38 \times 7{,}632{,}392 \qquad (4.56)$$

$$Y_4 + Y_5 + Y_6 + Y_7 + Y_8 - d_4^+ + d_4^- = 1.28 \times 12{,}348{,}981 \qquad (4.57)$$

$$\sum_{i=1}^{22} R_i^X X_i - \sum_{j=1}^{20} R_j^Y Y_j - d_5^+ + d_5^- = 30\% \times 17{,}327{,}046 + 653{,}116 \qquad (4.58)$$

$$X_1 - 0.01 \times 17{,}327{,}046 + d_6^- - d_6^+ = 0 \qquad (4.59)$$

$$X_2 - 0.004 \times 17{,}327{,}046 + d_7^- - d_7^+ = 0 \qquad (4.60)$$

$$X_3 - 0.14 \times 17{,}327{,}046 + d_8^- - d_8^+ = 0 \qquad (4.61)$$

$$X_{22} - 0.015 \times 17{,}327{,}046 + d_9^- - d_9^+ = 0 \qquad (4.62)$$

$$Y_4 + Y_5 + Y_6 + d_{10}^- - d_{10}^+ = 73.31\% \times 17{,}327{,}046 \qquad (4.63)$$

$$X_i \geq 0,\ Y_j \geq 0,\ d_k^+ \geq 0, d_k^- \geq 0, \text{ for all } i=1, 2, \ldots, 22, j=1, 2, \ldots, 20, k=1, 2, \ldots, 10 \qquad (4.64)$$

The objective function involves the minimization of the deviations d^+ and d^- from the target values of goals, where d^+ denotes the over-achievement of a goal and d^- the under-achievement. The deviations corresponding to different goals are weighted in the objective according to the significance of the goals. It should be mentioned that the above goal programming formulation is based on the version that gives first priority level to the solvency goal, second priority level to the liquidity goal and third priority level to the rest of the goals (version 1). More precisely, the selected weighted scheme assigns higher weight to under-achievement of the solvency goal (d_1^-), considering that it is achieved with a priority rank 3/2 higher than the priority rank that is imposed on the under achievement of the liquidity goal (d_2^-) and three times higher than the priority rank of the remaining goals ($(d_k^+ + d_k^-)$, $\forall k = 3,..,10$).

3. POST-OPTIMALITY

The solution of the above goal programming problem leads to an optimal solution, for which the corresponding optimal (minimum) value of the objective function is z^*. Once this optimal solution is obtained, a post-optimality stage takes place to investigate the sensitivity of the optimal solution. This is achieved through the investigation of the existence of sub-optimal solutions that correspond to objective function values lower than $z^* + k(z^*)$, where $k(z^*)$ is a small portion of the optimal solution z^*. In this case study $k(z^*)$ is considered equal to 5% of z^*.

This additional constraint is incorporated into the initial goal programming formulation and the new goal programming that is formed, is solved 42 times (once for each variable). Each of the 42 obtained solutions corresponds to the maximization of the asset and liability variables and determines the structure of the asset and liability data. In order to achieve a more efficient solution for the structure of the bank asset and liability, we take into account the estimation of the solution that was obtained after the post-optimality. Thus, the average of the 42 solutions obtained during the post optimality analysis is the maximum optimal solution of the asset and liability data and constitutes the final ALM solution.

4. INTEREST RATE SIMULATION ANALYSIS

Chapter 1 describes various risk management techniques, of which the scenario analysis is the most important. Having known, as already mentioned, that the interest rates of loans, deposits and bonds contribute significantly to the bank asset and liability management, it arises that the variability of the interest rate values constitutes a significant assigning factor of the decisions that the financial institutions take concerning the management of their assets and liabilities.

In the present study, the major unknown element in the above goal programming formulation which is of interest to the bank's managers is the return of the assets and liabilities used in the goal constraint (4.58). These involve the bonds' interest rates, the interest rates of the loans granted and the interest rates of the deposits to the bank. As already mentioned in Chapter 1, the changes of the interest rates affect the bank asset and liability management, while the right and successful management of the interest rate risk contribute to its profitability. To cope with the uncertainty on these parameters a scenario analysis approach is employed. This analysis involves the consideration of 2,500 scenarios on the aforementioned uncertain parameters. Monte Carlo simulation is used for the creation of the scenarios in the interest rate values in order to take into account the interest rate risk, as described in Chapter 3.

More precisely, initially, 50 scenarios are created for the interest rates of the deposits R_D of the bank. The deposit's interest rates range (approximately) between 3.5% and 7% and are considered as normally distributed random variables with appropriate mean 5.25% and standard deviation 0.55%. The choice of the values for these parameters of the normal distribution was realized based on the fact that the 99.7% of the observations that refer to a normal distributed random variable stands among the interval $[\mu-3\sigma, \mu+3\sigma]$ (Griffiths et al., 1992). Thus, the choice $\mu=5.25\%$ and $\sigma=0.55\%$, entails that the 99.7% of random scenarios that are developed during the simulation for the deposit interest rates will be among the interval [3.6%, 6.9%] which corresponds to the prespecified interval [3.5%, 7%].

For each deposit's interest rate scenario, 50 scenarios are generated for the bond's interest rates. The bond's interest rates are generated as normally distributed random variables, whose values range between 10% and 13%. Similarly to the deposits' interest rates, in order to guarantee that the bond's interest rates range among the pre-specified interval, it is assumed that they follow normal distribution with mean 11.5% and standard deviation 0.4%.

Besides the deposits' and bond's interest rates, the third form of interest rates, which is taken into account at the analysis refers to the loans' interest rates. Knowing that the banks determine their loans' interest rates at higher levels than those of the deposits, the loans' interest rates R_L, at the present study, are specified as follows:

$$R_L=R_D+S, \tag{4.65}$$

where S is the spread between the loans' interest rates R_L and the deposits' interest rates R_D. The spread ranges among 3% and 4% and is defined as normally distributed random variables with appropriate mean 3.5% and standard deviation 0.13%.

The interval width of the above values of interest rates was based on the empirical analysis of the historical data of past years to the interest rates' values, as shown in Table 4.2. It is worth mentioning that taking into account the values of the interest rates for the year 1999, as presented to Table 4.2, we observe a difference between the interest rates of deposits and the interest rates of loans of 6.5% to 7%. In this book, the spread among the interest rates of loans and the interest rates of deposits varies between 3% and 4%. It is a fact, that from one side small values of spread mean low profitability for the bank, as the margin profit between the loans and the deposits gets reduced. However, the reduction of the spread could be adopted by the bank during the marketing policy, aiming at attracting new customers for financing. This becomes obvious taking into account the small margins of reduction of the deposits' interest rates, which since 1999 have been to relative low levels and the comparative high margins of the loans' interest rates. Reducing the loans' interest rates at a higher degree than the reduction of the deposits' interest rates, leads to the reduction of the spread. The reduction of the loans' interest rates means optimal loan conditions for the bank's customers, which could lead to the increase of the customer basis. Thus, from this point of view it could be considered that the low spread that is adopted at the present simulation examines a "negative" scenario for the bank, according to which the bank adopts a policy of small margin profit aiming at the development of its customer basis to the sector of loans, at which there is increased competition to Greece during the last years.

Concluding, the distributions of the above variables of interest rates are presented as follows:

$R_D \sim N(5.25\%, 0.55\%)$

$R_B \sim N(11.5\%, 0.4\%)$

$Spread \sim N(3.5\%, 0.13\%)$

Table 4.2: Values of loans', deposits' and bonds' interest rates (Source: Alpha Bank-Interest rates 14/6/1996-18/12/2000)

	Loan interest rates	Deposit interest rates	Bond interest rates
14/6/1996	20.03%	9.56%	
1/9/1996	18.93%	9.56%	
19/5/1997	15.09%	6.79%	
17/11/1997	17.00%	8.38%	11.10%
24/12/1997	17.43%	8.38%	11.38%
16/3/1998	17.34%	7.63%	
23/3/1998	15.93%	7.10%	11.38%
2/11/1998	15.66%	7.10%	10.30%
18/1/1999	12.71%	6.52%	9.50%
20/12/1999	12.58%	6.08%	8.26%
31/1/2000	12.41%	4.91%	
14/3/2000	12.16%	4.85%	
24/4/2000	11.59%	5.19%	
10/7/2000	15.29%	4.83%	
11/9/2000	11.10%	3.71%	
27/11/2000	12.21%	2.86%	6.65%
18/12/2000	10.13%	2.35%	8.26%

Summarizing all the above in Table 4.3, the statistics for the bond, deposit and loan interest rates are described.

Table 4.3: Statistics for the bond, deposit and loan interest rates

	Bond interest rates	Deposit interest rates	Loan interest rates	Spread
Average value	11.53%	5.24%	8.74%	3.50%
Median	11.51%	5.28%	8.80%	3.5%
Standard deviation	0.40%	0.56%	0.58%	0.13%
Standard error	0.008%	0.011%	0.012%	0.003%
Min	10.29%	3.95%	7.10%	3.94%
Max	12.83%	6.11%	9.97%	3.07%

5. ANALYSIS OF RESULTS

The goal programming formulation (4.42)-(4.64) is solved for each of the 2,500 interest rate scenarios and 2,500 different solutions are obtained. Each of these solutions is then evaluated along all scenarios. This evaluation leads to the calculation of the expected present value (cf. constraint 4.58) and the corresponding risk of the expected return (standard deviation) among the solutions. This analysis leads to 207 non-dominating solutions in terms of the expected value and risk. Each of these solutions can be considered as ALM strategies that might be implemented during the next year. For a comparison the actual strategy (AS) that the bank followed during the year 20X0 is also considered along the same dimensions. Moreover, it should be mentioned at this point that the values of the results are presented in euro.

Table 4.4 presents the results (expected value, in euro) for AS and for 10 of the non-dominating solutions (ND_1, ND_2, ..., ND_{10}), which are similar to the actual strategy of the bank (in terms of the expected value and risk).

Table 4.4: Simulation analysis results based on version 1
(1st priority: solvency goal, 2nd priority: liquidity goal, 3rd priority: rest goals)

Solutions	Expected return	Standard deviation
AS	1,938,864.22	55,437.14
ND_1	2,130,981.42	55,004.64
ND_2	2,128,391.16	44,105.00
ND_3	2,121,575.94	43,862.38
ND_4	2,114,888.59	43,633.00
ND_5	2,113,781.52	43,515.20
ND_6	2,111,739.58	42,837.61
ND_7	2,107,943.52	42,560.53
ND_8	2,107,422.41	42,531.65
ND_9	2,104,687.38	42,462.88
ND_{10}	2,103,863.46	42,164.79

Taking into account the final values of all decision variables of the above 10 selected, non-dominating solutions, we calculate the maximum and the minimum value for each of the 42 decision variables, which are presented at Table 4.5. Table 4.6 presents the values of the most important decision variables for the 10 non-dominating solutions.

If we analyze the content of the above Tables 4.4, 4.5 and 4.6 and compare it with the actual results of the bank's financial statement we conclude that the accounts that refer to Cheques receivable and Deposits to the Bank of Greece (X_2, X_3) do not differ significantly from those of the actual strategy. The values of the variable X_2 range within an interval (70,642.61, 71,089.32) and those of X_3 range within the interval (2,786,565.37, 2,786,643.04), whereas the corresponding actual values are for the account Cheques receivable (X_2=79,080) and for the Deposits at the Bank of Greece (X_3=1,884,009). This is due to the restriction imposed, that the variables of fixed assets, cash, cheques receivable and deposits with the Bank of Greece should remain at the levels of previous years. Similarly, the actual value of the account Treasury bills and other securities issued by the Greek State (X_4) range within the interval that arises from the solution of the proposed asset liability management problem (6,976,945$\in$(164,344.93, 4,160,295.45)). The actual values of the accounts Other Treasury bills and securities (X_5) and Interbank deposits and loans repayable on demand (X_6) for the year 20X0 (X_5=198 and X_6=270) differ significantly from the solutions obtained through the proposed goal programming model ($X_5 \in$(2,352,984.57, 2,804,082.71) and $X_6 \in$(11,433.01, 113,105.91)). This is due to the fact that these variables determine the goal of asset and liability return, for which the interest rate scenarios were taken into account. Thus, changes the interest rates values and determination of these values on the basis of the bank managers' preferences could change the values of the above variables. The actual value of the account Other interbank deposits and loans (X_7=3,395,164) diverge and does not present significant difference from the width of the values interval of the goal programming model (3,502,953.98, 3,651,954.01). Similarly the actual value of the account Loans and advances to customers (X_8) diverges slightly upwards from the maximum value of the variable X_8, where the proposed interval width is (189,120.55, 7,679,796.41). The values of the accounts Loans and advances to customers maturing after one year (X_9) and Other receivables (X_{10}) that were obtained through the proposed goal programming model ($X_9 \in$(70,915.23, 128,609.86) and $X_{10} \in$(189,120.55, 7,679,605.40) differ significantly from the actual values (X_9=3,610,829 and X_{10}=19,373). This is due to the fact that several restrictions were imposed on the loans goals during the research. Moreover, the changes of the loans interest rates were taken into account and 2,500 interest rate scenarios were created. Similarly, the values of the accounts due to credit institutions (Y_1), credit institutions with agreed maturity (Y_2), commitments arising out of sale and repurchase transactions (Y_3), deposits repayable on demand (Y_4), saving deposits (Y_5), deposits with agreed maturing (Y_6), cheques and orders payable (Y_7), as well as the commitments arising out of sale and repurchase transactions (customer accounts) (Y_8) differ significantly from the actual values of the year 20X0. This is due to the fact that based on the bank data of past years the average growth rate of loans and deposits have been calculated and several constraints on the goals of loans and deposits were taken into account. Moreover, as it has already been mentioned during the formulation and development of the model, the asset and liability management is related to the uncertainty of

Table 4.5: **Maximum and minimum value for the 10 non dominated solutions based on version 1 (1st priority: solvency goal, 2nd priority: liquidity goal, 3rd priority: rest goals)**

Variables	*AS*	Min	Max
X_1	260,373	13,201.56	15,676.85
X_2	79,080	70,642.61	71,089.32
X_3	1,884,009	2,786,565.37	2,786.643.04
X_4	6,976,945	164,344.93	4,160,295.45
X_5	198	2,352,984.57	2,804,082.71
X_6	270	11,433.01	113,105.91
X_7	3,395,164	3,502,953.98	3,651,954.01
X_8	8,262,492	189,120.55	7,679,796.41
X_9	3,610,829	70,915.23	128,609.86
X_{10}	19,373	189,120.55	7,679,605.40
X_{11}	386,035	164,344.93	4,289,450.72
X_{12}	421,372	74,276.15	129,916.68
X_{13}	136,658	18,263.14	31,706.51
X_{14}	11,772	111,433.01	113,105.91
X_{15}	1,254,839	111,433.01	113,105.91
X_{16}	255,291	95,600.37	108,722.77
X_{17}	5,413	95,600.37	108,722.77
X_{18}	322,504	95,600.37	108,722.77
X_{19}	11,110	95,600.37	108,722.77
X_{20}	61,914	95,600.37	108,722.77

X_{21}	68,504	95,600.37	108,722.77
X_{22}	963,500	247,943.67	252,701,36
Y_1	851,489	24,298.12	27,695.33
Y_2	3,205,814	23,602.78	26,304.01
Y_3	986,569	67,995.76	168,129.37
Y_4	3,070,110	12,087,319.52	12,087,491.95
Y_5	6,712,749	305,308.60	309,578.33
Y_6	5,637,741	305,371.79	310,495.26
Y_7	189,980	3,009,223.65	3,018,212.77
Y_8	4,515,094	83,276.81	83,436.93
Y_9	147,582	27,150.91	31,460.65
Y_{10}	142,256	27,150.91	31,460.65
Y_{11}	146,193	27,150.91	31,460.65
Y_{12}	132,540	27,150.91	31,460.65
Y_{13}	4,383	27,150.91	31,460.65
Y_{14}	38,160	27,150.91	31,460.65
Y_{15}	132,510	27,150.91	31,460.65
Y_{16}	8,872	27,150.91	31,460.65
Y_{17}	62,925	27,150.91	31,460.65
Y_{18}	275,000	27,150.91	31,460.65
Y_{19}	744,577	5,024,162.69	5,108,500.67
Y_{20}	49,466	526,970.02	528,990.49

Table 4.5: Maximum and minimum value for the 10 non dominated solutions based on version 1 (1st priority: solvency goal, 2nd priority: liquidity goal, 3rd priority: rest goals) (continued)

Table 4.6: Results of the significant decision variables for the 10 non-dominated solutions based on version 1 (1st priority: solvency goal, 2nd priority: liquidity goal, 3rd priority: rest goals)

Variables	AS	ND_1	ND_2	ND_3	ND_4	ND_5	ND_6	ND_7	ND_8	ND_9	ND_{10}
X_2	79,080	71,072	70,988	70,929	70,993	70,932	71,012	70,642	70,972	71,089	71,038
X_3	1,884,009	2,786,630	2,786,624	2,786,629	2,786,643	2,786,617	2,786,627	2,786,611	2,786,565	2,786,583	2,786,606
X_4	6,976,945	4,160,295	164,345	164,345	164,345	3,881,027	164,345	164,345	164,345	164,345	3,938,438
X_5	198	2,479,986	2,577,139	2,523,555	2,352,984	2,746,119	2,666,776	2,567,874	2,804,082	2,683,799	2,650,145
X_6	270	112,634	111,835	111,485	112,278	111,433	112,102	113,105	111,503	112,187	112,089
X_8	8,262,492	7,676,225	189,120	7,674,591	7,679,796	189,120	7,676,817	7,672,280	189,120	7,621,932	189,120
X_9	3,610,829	74,451	76,322	76,081	70,915	79,104	73,850	78,341	78,065	128,609	71,002
X_{10}	19,373	189,120	7,674,334	189,120	189,120	7,671,534	189,120	189,120	7,672,425	189,120	7,679,605
Y_1	851,489	25,462	25,536	24,972	24,298	27,104	26,513	26,258	27,695	26,400	26,462
Y_2	3,205,814	24,850	24,414	23,847	23,602	25,631	25,689	25,864	26,304	25,632	25,392
Y_3	986,569	70,293	157,236	155,150	67,995	166,535	160,380	72,932	168,129	163,589	157,737
Y_4	3,070,110	12,087,411	12,087,446	12,087,492	12,087,482	12,087,483	12,087,439	12,087,319	12,087,437	12,087,346	12,087,404
Y_5	6,712,749	306,458	307,060	306,120	305,308	305,769	305,344	307,604	308,345	308,536	309,578
Y_6	5,637,741	307,115	307,286	306,201	307,572	305,371	307,608	307,733	306,873	307,275	310,495
Y_7	189,980	3,015,859	3,015,045	3,017,097	3,016,617	3,018,213	3,016,461	3,013,993	3,013,842	3,013,359	3,009,223
Y_8	4,515,094	83,352	83,319	83,276	83,286	83,285	83,325	83,436	83,327	83,411	83,358
Y_{19}	744,577	5,084,589	5,066,385	5,076,545	5,108,501	5,034,163	5,048,552	5,068,722	5,024,163	5,047,555	5,052,445
Y_{20}	49,466	527,134	528,695	528,555	526,970	528,853	528,836	527,311	528,990	528,947	528,878

the risk management and especially of the interest rate risk. The simulation analysis through the scenario generation becomes essential for the values of deposits', loans' and bonds' interest rates. The consideration of these scenarios contributes to the choice for the bank's optimal solution. In case different interest rate scenarios are taken into account, different optimal solutions may arise.

5.1 Sensitivity analysis to the priorities of goals

A sensitivity analysis took place exploring the effects of alterations in the order of goal priorities. According to the first version, the first priority level is given to the solvency goal, the second priority level is given to the liquidity goal and the third to the rest goals including the loans and deposits goals. At the second version, the first priority level is given to the liquidity goal, the second priority level to the solvency goal and the third priority level to the rest goals (loan goal, deposit goal, etc). Reordering the priority levels (version 2) produces the optimal solution in Table 4.7 for the actual strategy and for 10 of the non-dominating solutions ($ND'_1, ND'_2, \ldots, ND'_{10}$).

Table 4.7: Simulation analysis results based on version 2
(1st priority: liquidity goal, 2nd priority: solvency goal, 3rd priority: rest goals)

Solutions	Expected return	Standard deviation
AS	1,938,864.22	55,437.14
ND'_1	2,130,947.93	55,007.54
ND'_2	2,128,422.47	44,103.41
ND'_3	2,121,607.22	43,859.93
ND'_4	2,113,233.20	43,582.24
ND'_5	2,112,101.98	43,463.85
ND'_6	2,111,768.84	42,835.52
ND'_7	2,107,974.04	42,558.45
ND'_8	2,107,453.12	42,529.57
ND'_9	2,104,718.17	42,460.81
ND'_{10}	2,101,786.31	42,108.90

Table 4.8 presents the minimum and maximum value of the final values for each of the decision variables, which arise from the pre-specified 10 non-dominated solution for the second version, while Table 4.9 presents the values of the most significant decision variables for the 10 non-dominated solutions.

If we analyze the content of Table 4.8, we conclude that the results are similar to the analysis of Table 4.5, which concerns the results of the decision variables of the first version. Similarly, at the second version, the values of the accounts due to credit institutions (Y_1), credit institutions with agreed maturity (Y_2), commitments arising out of sale and repurchase transactions (Y_3), deposits repayable on demand (Y_4), saving deposits (Y_5), deposits with agreed maturity (Y_6), cheques and orders payable (Y_7), as well as commitments arising from sale and repurchase transactions (customer amounts) (Y_8) differ significantly from the actual values of the year 20X0. Thus, the actual values of the accounts loans and advances to customers maturing within one year (X_9) and other receivables (X_{10}) present significant divergence from the solutions obtained by the goal programming model at the present study. This indicates that the main source of cash flows for the bank is the accounts of loans and deposits. As already mentioned in the previous section, the consideration of scenarios towards the values of deposits', loans' and bonds' interest rates contributes to the determination of an optimal solution for the bank. Thus, a different scenario consideration contributes to different results concerning the structure of asset and liability.

If we analyze the content of Table 4.6 and Table 4.9 we conclude that the comparison of the solution of the second version of the problem with the initial one indicates that the optimal values of the variables change marginally. This is due to the fact that the initial solution does not diverge from the values of loan and deposit goals. Reordering the priority levels, by giving first priority level to the liquidity goal and second priority level to the

Table 4.8: Maximum and minimum value for the 10 non dominated solutions based on version 2 (1st priority: liquidity goal, 2nd priority: solvency goal, 3rd priority: rest goals)

Variables	*AS*	Min	Max
X_1	260,373	13,201.56	15,676.85
X_2	79,080	70,642.61	71,089.32
X_3	1,884,009	2,786,231.69	2,786,330.22
X_4	6,976,945	164,344.93	4,159,508.52
X_5	198	2,352,984.57	2,804,082.71
X_6	270	111,433.01	113,105.91
X_7	3,395,164	3,505,012.90	3,653,999.84
X_8	8,262,492	189,120.55	7,678.905.09
X_9	3,610,829	70,915.23	141,399.08
X_{10}	19,373	189,120.55	7,678,671.68
X_{11}	386,035	164,344.93	4,288,676.16
X_{12}	421,372	74,276.15	142,579.30
X_{13}	136,658	18,263.14	34,775.08
X_{14}	11,772	111,433.01	113,105.91
X_{15}	1,254,839	111,433.01	113,105.91
X_{16}	255,291	95,600.37	108,722.77
X_{17}	5,413	95,600.37	108,722.77
X_{18}	322,504	95,600.37	108,722.77
X_{19}	11,110	95,600.37	108,722.77
X_{20}	61,914	95,600.37	108,722.77
X_{21}	68,504	95,600.37	108,722.77
X_{22}	963,500	247,943.67	252,701.36
Y_1	851,489	24,298.12	27,695.33
Y_2	3,205,814	23,602.78	26,304.01
Y_3	986,569	67,995.76	168,129.37
Y_4	3,070,110	12,087,319.52	12,087,491.95
Y_5	6,712749	305,308.60	309,578.33
Y_6	5,637,741	305,371.79	310,495.26
Y_7	189,980	3,007,365.55	3,016,378.87
Y_8	4,515,094	83,276.81	83,436.93
Y_9	147,582	27,150.91	31,460.65
Y_{10}	142,256	27,150.91	31,460.65
Y_{11}	146,193	27,150.91	31,460.65
Y_{12}	132,540	27,150.91	31,460.65

Y_{13}	4,383	27,150.91	31,460.65
Y_{14}	38,160	27,150.91	31,460.65
Y_{15}	132,510	27,150.91	31,460.65
Y_{16}	8,872	27,150.91	31,460.65
Y_{17}	62,925	27,150.91	31,460.65
Y_{18}	275,000	27,150.91	31,460.65
Y_{19}	744,577	5,026,054.62	5,110,274.40
Y_{20}	49,466	526,970.02	528,990.49

Table 4.8: **Maximum and minimum value for the 10 non dominated solutions based on version 2 (1st priority: liquidity goal, 2nd priority: solvency goal, 3rd priority: rest goals) (contiued)**

Table 4.9: **Results of the significant decision analysis for the 10 non-dominated solutions based on version 2 (1st priority: liquidity goal, 2nd priority: solvency goal, 3rd priority: rest goals)**

Variables	*AS*	ND'_1	ND'_2	ND'_3	ND'_4	ND'_5	ND'_6	ND'_7	ND'_8	ND'_9	ND'_{10}
X_2	79,080	71,072	70,988	70,929	70,993	70,932	71,012	70,642	70,972	71,089	71,038
X_3	1,884,009	2,786,312	2,786,303	2,786,311	2,786,330	2,786,294	2,786,308	2,786,286	2,786,232	2,786,247	2,786,279
X_4	6,976,945	4,159,508	164,345	164,345	164,345	3,877,158	164,345	164,345	164,345	164,345	3,925,191
X_5	198	2,479,986	2,577,140	2,523,555	2,352,985	2,746,119	2,666,776	2,567,875	2,804,082	2,683,799	2,650,145
X_6	270	112,633	111,835	111,485	112,278	111,433	112,102	113,105	111,503	112,187	112,089
X_8	8,262,492	7,675,319	189,120	7,673,683	7,678,905	189,120	7,675,907	7,671,353	189,120	7,608,183	189,120
X_9	3,610,829	74,451	76,322	76,081	70,915	79,104	73,850	78,341	78,065	141,399	71,002
X_{10}	19,373	189,120	7,673,420	189,120	189,120	7,670,613	189,120	189,120	7,671,474	189,120	7,678,671
Y_1	851,489	25,462	25,536	24,972	24,298	27,104	26,513	26,258	27,695	26,400	26,462
Y_2	3,205,814	24,851	24,414	23,847	23,602	25,631	25,689	25,864	26,304	25,632	25,393
Y_3	986,569	70,294	157,236	155,150	67,996	16,.535	160,380	72,932	168,129	163,590	157,737
Y_4	3,070,110	12,087,411	12,087,446	12,087,492	12,087,482	12,087,483	12,087,439	12,087,319	12,087,437	12,087,346	12,087,404
Y_5	6,712,749	306,458	307,060	306,120	305,308	305,769	305,344	307,604	308,345	308,536	309,578
Y_6	5,637,741	307,115	307,286	306,201	307,572	305,371	307,608	307,733	306,873	307,275	310,495
Y_7	189,980	3,014,057	3,013,286	3,015,289	3,014,843	3,016,378	3,014,650	3,012,147	3,011,950	3,011,448	3,007,365
Y_8	4,515,094	83,351	83,319	83,276	83,286	83,285	83,325	83,436	83,327	83,411	83,358
Y_{19}	744,577	5,086,391	5,068,204	5,078,352	5,110,274	5,035,996	5,050,362	5,070,567	5,026,054	5,049,465	5,054,303
Y_{20}	49,466	527,134	528,695	528,555	526,970	528,853	528,836	527,311	528,990	528,947	528,878

solvency goal, does not affect the results (version 2).

It is observed that the values of the account "Cheques receivable" (X_2) and "Deposits at the Bank of Greece" (X_3) do not differ significantly form both version. This is due to the constraint that was imposed to the variables of fixed assets, cash, cheques receivable and deposits at the bank of Greece that remain at the previous year's levels. The value of the account "Treasury bills and other securities issued by the Greek State" (X_4) of the solutions ND_1, ND_5, ND_{10} of the first version and the solution ND'_1, ND'_5, ND'_{10} of the second version does not differ significantly from the results of the actual strategy. Similarly, the values of the account "Loans and advances to customers maturing within one year" (X_8) do not present significant differences from the actual results besides the values based on the solutions ND_2, ND_5, ND_8, ND_{10} of the first version and the solutions ND'_2, ND'_5, ND'_8, ND'_{10} of the second version. The above results indicate that the main source of cash flows for a commercial bank is the accounts of loans and deposits, as already mentioned in the previous sections. The values of the accounts " Loans and advances to customers maturing after one year" (X_9), "Other receivables" (X_{10}), "Due to credit institutions" (Y_1), "Due to credit institutions with agreed maturity" (Y_2), "Commitments arising out of sale and repurchase transactions" (Y_3), "Deposits repayable on demand" (Y_4), "Saving deposits" (Y_5), "Deposits with agreed maturity" (Y_6), "Cheques and orders payable" (Y_7) and "Commitments arising out of sale and repurchase transactions (customer amounts)" (Y_8) differ significantly from the results of the actual strategy to both versions. It is observed that the above variables concern those of loans and deposits. The most important difference that arises is due to the fact that specific constraints should be imposed on the goals of loans and deposits during the research. Based on the historical data of the bank financial statements, it is assumed that the total deposits and loans are not expected to increase by more than 28% and 38% respectively over the previous year's levels. Moreover, since the asset liability management is related to uncertainty, the simulation analysis becomes essential and thus the consideration of scenarios about the loans' and deposits' interest rates contributes to the choice of the optimal solution for the bank.

It should be mentioned that although there are significant differences among the values of the basic account of the balance sheet, such as the accounts of deposits and loans, the structure of the balance sheet does not present significant difference to the 10 non-dominated solutions. It is worth mentioning that the accounts of deposits and loans are considered to be significant since they constitute the major source of cash flows for the bank.

5.2 Forecasting analysis

The uncertainty and the risk that exist due to the liberation of the financial markets, the globalization, the technological innovations and the eminent competition among the firms and the financial institutions, force the banks to develop long term strategies in order to proceed to the most efficient financial risk management and the optimal structure of their assets and liabilities.

During the development of a bank asset management model at the present study, taking into account the values of the assets and liabilities that rose at 20X0, a forecast and determination of the assets and liabilities for the year 20X1 will take place. Thus, the bank managers could attend the financial situation of the bank and proceed to long term strategies of asset liability management taking into account the economic progress and various alternatives changes to the market parameters.

The non-dominating solution ND_1 of the first version is selected as the best solution based on the criterion of the expected return. It is reminded that based on the first version, first priority level is given to the solvency goal and second priority level to the liquidity goal. Taking into account the final values of the decision variables that are based on the solution ND_1 a series of constraints and goals is modulated as analyzed in section 2 of the present chapter.

More specifically, the goal programming problem based on the decision variables of the non-dominated solution ND_1 is modulated as follows:

$$\text{Min } z = \sum_{k=3}^{10} d_k^+ + \sum_{k=3}^{10} d_k^- + 2d_2^- + 3d_1^- \qquad (4.66)$$

subject to the constraints

$$X_8 + X_9 + X_{10} \geq 7.939.797 \quad (4.67)$$

$$X_8 + X_9 + X_{10} \leq 1{,}42 \times 7.939.797 \quad (4.68)$$

$$Y_4 + Y_5 + Y_6 + Y_7 + Y_8 \geq 15.800.196 \quad (4.69)$$

$$Y_4 + Y_5 + Y_6 + Y_7 + Y_8 \leq 1{,}35 \times 15.800.196 \quad (4.70)$$

$$Y_{19} \geq 5.084.590 \quad (4.71)$$

$$Y_{20} \geq 1{,}89\% \times \sum_{i=1}^{22} X_i \quad (4.72)$$

$$Y_4 + Y_5 + Y_6 + Y_7 + Y_8 - 1{,}94(X_8 + X_9 + X_{10}) = 0 \quad (4.73)$$

$$Y_4 + Y_5 + Y_6 + Y_7 + Y_8 - 2{,}33(X_4 + X_5 + X_{11} + X_{12} + X_{13}) = 0 \quad (4.74)$$

$$Y_4 + Y_5 + Y_6 + Y_7 + Y_8 - 6{,}51 X_3 = 0 \quad (4.75)$$

$$\sum_{i=1}^{22} X_i - \sum_{j=1}^{20} Y_j = 1.162.840 \quad (4.76)$$

$$\sum_{i=1}^{22} X_i \leq 1{,}35 \times 22.454.795 \quad (4.77)$$

$$\begin{aligned} &Y_{19} - 0{,}2469 Y_{20} - 0{,}2 X_4 - 0{,}5 X_8 - 0{,}7 X_9 - \\ &0{,}5 X_{10} - 0{,}2 X_{11} - 0{,}4 X_{12} - X_{13} - d_1^+ + d_1^- = 8\% \end{aligned} \quad (4.78)$$

$$\sum_{i=1}^{21} X_i - 0{,}6 \sum_{j=1}^{18} Y_j - d_2^+ + d_2^- = 0 \quad (4.79)$$

$$X_8 + X_9 + X_{10} - d_3^+ + d_3^- = 1{,}42 \times 7.939.797 \quad (4.80)$$

$$Y_4 + Y_5 + Y_6 + Y_7 + Y_8 - d_4^+ + d_4^- = 1{,}35 \times 15.800.196 \quad (4.81)$$

$$\sum_{i=1}^{22} R_i^X X_i - \sum_{j=1}^{20} R_j^Y Y_j - d_5^+ + d_5^- = 35\% \times 22.454.795 + 1.162.840 \quad (4.82)$$

$$X_1 - 0{,}01 \times 22.454.795 + d_6^- - d_6^+ = 0 \tag{4.83}$$

$$X_2 - 0{,}004 \times 22.454.795 + d_7^- - d_7^+ = 0 \tag{4.84}$$

$$X_3 - 0{,}13 \times 22.454.795 + d_8^- - d_8^+ = 0 \tag{4.85}$$

$$X_{22} - 0{,}018 \times 22.454.795 + d_9^- - d_9^+ = 0 \tag{4.86}$$

$$Y_4 + Y_5 + Y_6 + d_{10}^- - d_{10}^+ = 72{,}52\% \times 22.454.795 \tag{4.87}$$

$$X_i \geq 0,\ Y_j \geq 0,\ d_k^+ \geq 0, d_k^- \geq 0, \text{ for all } i=1, 2, \ldots, 22, j=1, 2, \ldots, 20, k=1, 2, \ldots, 10 \tag{4.88}$$

Following the development steps of the asset liability management model, as they are described in the flowchart of Chapter 3, realizing the post optimality stage and the approach of the scenario analysis for the bonds', loans' and deposits' interest rates, as they are described in the previous section, the above goal programming model is solved for each of the 2,500 interest rate scenarios and 2,500 different solutions are obtained.

Similarly, each of these solutions is evaluated based on the expected return and the corresponding risk (standard deviation). Thus, 207 non-dominating solutions arise, based on the expected return and the risk.

Table 4.10 presents the results of 10 from the non-dominating forecasted solutions (NDf_1, NDf_2, ..., NDf_{10}) for the year 20X1.

Table 4.10: Simulation analysis results aiming at the forecasting analysis (version 1)

Soltions	Expected return	Standard deviation
NDf_1	2,896,534	59,572
NDf_2	2,896,438	59,542
NDf_3	2,893,092	59,351
NDf_4	2,892,541	59,325
NDf_5	2,886,032	59,167
NDf_6	2,873,654	59,059
NDf_7	2,870,199	58,995
NDf_8	2,868,159	58,893
NDf_9	2,868,100	58,708
NDf_{10}	2,867,992	58,629

As already mentioned in the previous section Table 4.11 presents the minimum and maximum value of the final values for each of the decision variables that are based on the pre-specified 10 non-dominating solutions, while Table 4.12 presents the values of the most significant decision variables for the 10 non-dominated solutions for version 1.

If we analyze the content of Table 4.12, we observe that the values of the variables that concern the partial accounts differ from most of the solutions and especially from the most basic categories of the balance sheet accounts. The account "Treasury bills and other securities issued by the Greek State" (X_4) does not present significant difference among the solutions NDf1, NDf_6, NDf_7, NDf_8 and NDf_{10}. Similarly the results of the account "Loans and advances to customers maturing within one year" (X_8) which are based on the solutions NDf_1, NDf_4, NDf_8, NDf_9, NDf_{10} present significant differences from the solutions NDf_2, NDf_3, NDf_5, NDf_6, NDf_7. The value of the variable that concerns the account "Due to credit institutions" (Y_1) and is based on the solution NDf_2 presents significant differences from the rest of the solutions.

Table 4.11: Maximum and minimum value for the 10 non-dominated solutions of the forecasting analysis based on version 1

Variables	Min value	Max value
X_1	219,201.57	219,201.57
X_2	95,968.65	96,365.41
X_3	3,272,590.25	3,274,116.88
X_4	217,053.57	6,667,365.77
X_5	2,144,449.24	2,561,763.26
X_6	89,924.71	110,222.24
X_7	5,121,494.70	5,275,951.13
X_8	261,785,28	10,654,677.08
X_9	66,425,09	196,101.49
X_{10}	261,785.28	10,657,164.62
X_{11}	217,009.04	6,619,529.65
X_{12}	68,911.38	422,995.06
X_{13}	18,475.76	45,758.12
X_{14}	142,006.16	150,746.84
X_{15}	142,006.16	150,746.84

X_{16}	89,924.71	110,222.24
X_{17}	89,924.71	110,222.24
X_{18}	89,924.71	110,222.24
X_{19}	89,924.71	110,222.24
X_{20}	89,924.71	110,222.24
X_{21}	89,924.71	110,222.24
X_{22}	403,198.12	405,756.11
Y_1	14,567.37	29,870.20
Y_2	13,230.04	14,576.10
Y_3	14,567.37	15,757.26
Y_4	15,877,572.88	15,983,679.42
Y_5	144,941.77	191,956.71
Y_6	144,267.18	205,988.68
Y_7	4,900,201.65	4,910,139.93
Y_8	128,725.65	128,978.09
Y_9	14,567.37	15,757.26
Y_{10}	14,567.37	15,757.26
Y_{11}	14,567.37	15,757.26
Y_{12}	14,567.37	15,757.26
Y_{13}	14,567.37	15,757.26
Y_{14}	14,567.37	15,757.26
Y_{15}	14,567.37	15,757.26
Y_{16}	14,567.37	15,757.26
Y_{17}	14,567.37	15,757.26
Y_{18}	14,567.37	15,757.26
Y_{19}	7,058,561.87	7,085,197.41
Y_{20}	594,806.58	595,860.67

Table 4.11: Maximum and minimum value for the 10 non-dominated solutions of the forecasting analysis based on version 1 (continued)

Table 4.12: Results of the significant decision variables of the forecasting analysis (version 1)

Variables	NDf_1	NDf_2	NDf_3	NDf_4	NDf_5	NDf_6	NDf_7	NDf_8	NDf_9	NDf_{10}
X_2	95,968	96,149	96,230	96,120	96,081	96365	96,178	96,139	96,076	96,113
X_3	3,273,417	3,273,695	3,272,961	3,274,101	3,273,655	3,273,043	3,272,590	3,273,325	3,274,116	3,273,675
X_4	6,633,069	217,176	217,053	217,265	217,325	6,033,869	6,667,365	5,980,567	217,246	6,044,984
X_5	2,181,982	2,219,091	2,198,297	2,554,929	2,231,725	2,561,763	2,144,449	2,506,203	2,244,356	2,477,826
X_6	89,925	98,063	110,222	92,189	99,415	110,166	106,496	103,849	91,302	98,107
X_8	261,785	10,654,677	10,647,646	26,.785	10,534,921	10,525,366	10,651,809	261,785	261,785	261,785
X_9	72,636	68,980	73,546	72,974	188,601	196,101	68,138	73,395	69,811	66,425
X_{10}	10,650,086	261,785	26,.785	10,652,045	261,785	261,785	261,785	10,649,019	10,655,259	10,657,164
Y_1	14,567	29,870	14,994	15,335	15,012	15,757	14,576	14,988	15,004	14,962
Y_2	13,230	13,722	13,917	13,913	13,815	14,576	13,519	13,808	13,599	13,673
Y_3	14,567	14,935	14,994	15,335	15,012	15,757	14,576	14,988	15,004	14,962
Y_4	15,983,679	15,963,902	15,937,113	15,941,457	15,932,001	15,935,917	15,945,389	15,901,941	15,948,969	15,877,572
Y_5	147,688	144,941	168,428	175,149	178,946	169,009	152,584	186,751	157,243	191,956
Y_6	144,267	166,718	169,870	158,957	164,499	170,456	177,551	186,734	169,399	205,988
Y_7	4,905,586	4,907,397	4,0902,617	4,910,041	4,907,138	4,903,151	4,900,201	4,904,988	4,910,140	4,907,266
Y_8	128,725	128,798	128,947	128,796	128,914	128,978	128,835	128,932	128,748	128,842
Y_{19}	7,080,213	7,058,561	7,077,368	7,065,855	7,072,731	7,067,023	7,085,197	7,075,177	7,070,050	7,073,347
Y_{20}	594,806	595,125	595,400	595,338	595,357	595,860	594,987	595,373	595,081	595,213

A sensitivity analysis takes place examining the effect of alterations to the priority order of the goals, as it is mentioned in the previous section. In the second version first priority level is given to the liquidity goal, second to the solvency goal and third to the rest of the goals.

Table 4.13 presents the results for 10 of the non-dominated forecasting solutions (NDf_1', NDf_2', …, NDf_{10}') for the year 20X1, based on version 2, while Table 4.12 presents the minimum and maximum value of the final values of all the decision variables based on the 10 non-dominated pre-specified solutions of version 2. Finally, Table 4.15 presents the values of the most significant decision variables for the 10 non-dominating solutions of version 2.

Table 4.13: Simulation analysis results aiming at the forecasting analysis (version 2)

Solutions	Expected return	Standard deviation
NDf_1'	2,896,451	59,630
NDf_2'	2,896,313	59,596
NDf_3'	2,892,942	59,408
NDf_4'	2,892,381	59,382
NDf_5'	2,885,883	59,220
NDf_6'	2,873,504	59,121
NDf_7'	2,870,048	59,054
NDf_8'	2,876,998	58,951
NDf_9'	2,867,945	58,763
NDf_{10}'	2,867,816	58,684

Similarly to the analysis of the content of Table 4.12 and the results of Table 4.15, we observe differences among the solutions of specific accounts. The comparison of the solution of version 2 (Table 4.15) of the problem with version 1 (Table 4.12) indicates that the optimal values of the variables change marginally. This is due to the fact that more importance is given to the priority levels of the liquidity and solvency goals and they do not diverge from the values of the constraints.

Concluding, based on the financial statement of the previous economic year, the preferences and the goals of the bank managers, the development of the above goal programming problem contributes to the determination of the asset and liability data for the year 20X0 and then for the year 20X1. It is observed that the changes to the order of the priority levels of the liquidity and solvency goals do not affect significantly the results. Taking into account all the above, the observations that were realized and the policy that was proposed during the development of the goal programming model, several standards of strategy are proposed for the bank contributing to the decision taking for the implementation of the above results.

Table 4.14: Maximum and minimum value for the 10 non-dominated solutions of the forecasting analysis based on version 2.

Variables	Min value	Max value
X_1	219,201.57	219.201,57
X_2	95,968.65	96.365,14
X_3	3,273,087.56	3.274.617,19
X_4	217,841	6.692.932,39
X_5	2,119,529.73	2.561.763,26
X_6	89,924.71	110.222,24
X_7	5,117,811.52	5.272.362,91
X_8	261,785.28	10.656.425,68
X_9	66,425.09	196.101,49
X_{10}	261,785.28	10.658.941,22
X_{11}	217,743.96	6.647.037,69
X_{12}	68,911.38	422.995,06
X_{13}	18,475.76	45.758,12
X_{14}	142,006.16	150.746,84
X_{15}	142,006.16	150.746,84
X_{16}	89,924.71	110.222,24
X_{17}	89,924.71	110.222,24
X_{18}	89,924.71	110.222,24
X_{19}	89,924.71	110.222,24
X_{20}	89,924.71	110.222,24
X_{21}	89,924.71	110.222,24
X_{22}	403,198.12	405.756,11
Y_1	14,567.37	29.870,20
Y_2	13,230.04	14.576,10
Y_3	14,567.37	15.757,26

Y_4	15,877,572.88	15.983.679,42
Y_5	144,941.77	191.956,71
Y_6	144,267.18	205.988,68
Y_7	4,903,439.08	4.913.396,94
Y_8	128,725.65	128.978,09
Y_9	14,567.37	15.757,26
Y_{10}	14,567.37	15.757,26
Y_{11}	14,567.37	15.757,26
Y_{12}	14,567.37	15.757,26
Y_{13}	14,567.37	15.757,26
Y_{14}	14,567.37	15.757,26
Y_{15}	14,567.37	15.757,26
Y_{16}	14,567.37	15.757,26
Y_{17}	14,567.37	15.757,26
Y_{18}	14,567.37	15.757,26
Y_{19}	7,055,169.59	7.081.959,90
Y_{20}	594,806.58	595.860,67

Table 4.14: Maximum and minimum value for the 10 non-dominated solutions of the forecasting analysis based on version 2 (continued)

Table 4.15: Results of the significant decision variables of the forecasting analysis (version 2)

Variables	NDf_1^*	NDf_2^*	NDf_3^*	NDf_4^*	NDf_5^*	NDf_6^*	NDf_7^*	NDf_8^*	NDf_9^*	NDf_{10}^*
X_2	95,968	96,149	96,230	96,120	96,081	96,365	96,178	96,139	96,076	96,113
X_3	3,273,919	3,274,216	3,273,471	3,274,595	3,274,115	3,273,576	3,273,087	3,273,841	3,274,617	3,274,205
X_4	6,660,727	217,967	217,841	217,967	217,967	6,034,522	6,692,932	5,981,205	217,967	6,045,641
X_5	2,154,991	2,192,249	2,171,467	2,554,929	2,206,438	2,561,763	2,119,529	2,506,203	2,218,225	2,477,826
X_6	89,925	98,063	110,222	92,189	99,415	110,166	106,496	103,849	91,302	98,107
X_8	261,785	10,656,425	10,649,356	261,785	10,536,463	10,527,156	10,653,478	261,785	261,785	261,785
X_9	72,636	68,980	73,546	72,974	188,601	196,101	68,138	73,395	69,811	66,425
X_{10}	10,651,770	261,785	261,785	10,653,701	261,785	261,785	261,785	10,650,750	10,656,938	10,658,941
Y_1	14,567	29,870	14,994	15,335	15,012	15,757	14,576	14,988	15,004	14,962
Y_2	13,230	13,722	13,917	13,913	13,815	14,576	13,519	13,808	13,599	13,673
Y_3	14,567	14,935	14,994	15,335	15,012	15,757	14,576	14,988	15,004	14,962
Y_4	15,983,679	15,963,902	15,937,113	15,941,457	15,932,000	15,935,917	15,945,389	15,901,941	15,948,969	15,877,572
Y_5	147,688	144,941	168,429	175,149	178,946	169,009	152,584	186,751	157,243	191,956
Y_6	144,267	166,718	169,870	158,957	164,499	170,456	177,551	186,734	169,399	205,988
Y_7	4,908,853	4,910,789	4,905,935	4,913,253	4,910,128	4,906,622	4,903,439	4,908,345	4,913,396	4,910,713
Y_8	128,725	128,798	128,947	128,796	128,914	128,978	128,835	128,932	128,748	128,842
Y_{19}	7,076,945	7,055,169	7,074,051	7,062,643	7,069,741	7,063,552	7,081,959	7,071,820	7,066,793	7,069,900
Y_{20}	594,806	595,125	595,400	595,338	595,357	595,860	594,987	595,373	595,081	595,213

6. POLICY AND STRATEGY STANDARDS OF THE BANKS

During recent years, the most significant institutional and structural changes that were realized in the banking field, the progress of technology and the liberalization of the financial markets increased the instability and the risks. The banks, which are subject to the changes raised due to the new economic and monetary environment with the creation of the united monetary system, enlarged their network in order to ameliorate their operating efficiency. While until the beginning of the decade of 2000, the income has increased significantly due to the financial effects that were achieved during the divergence of the Greek interest rates from the lowest European interest rates and of the favorable stock market situation, during recent years we observe a loss of significant income from the radically negative developments of the Greek financial market. The intense competitive conditions that stand to most markets necessitate the Greek banking institutions changing policy aiming at the increase of their operating income and thus of their profitability.

During the restructuring of the operations of the banking environment, the development of the above goal programming model provides the possibility to the bank managers to apply their long term strategies of asset liability management in order to succeed the optimal and efficient management policy.

Taking into account the policy that was followed for the development of the proposed model, in the present study, through the formulation of constraints and goals, as they are described analytically in the above sections, interesting conclusions are obtained. The analysis of the results obtained for the year 20X0 and their comparison with the actual strategy that was followed during this year, indicated that there are significant differences from the variables of the accounts of demands and liabilities towards the customers and the financial institutions. It is a fact, that in a commercial bank, the largest percentage of income arises from the deposits and the loans and for this reason, the pre-specified accounts are assumed to be the most significant to the configuration of the bank balance sheet. The results of the present research imply several strategies that the bank could possibly follow in order to modulate its balance sheet to accomplish an optimal asset liability management.

√ The promotion of new banking products, as well as the inspection of the products which are disposed to the investors and the borrowers will contribute to attracting a large number of customers and thus to the optimal banking investment and financing of large firms. The increasing demands of the customers for more complicated and flexible financial products impose on the banking institutions the invention of different sources of their competitive advantage and are directed towards more specialized markets which will guarantee their success. Thus, the high quality of products and services of a bank as well as their diversification against those produced by another competitor and the speed of their entering the market are of great importance for each financial institution (Kochan and Dyer, 1993).

√ Besides the quality of new financial products, another factor which contributes to the increase of the member of customers and receives great importance in the competitive environment is the quality of customer service, which presupposes the best staff training. The importance of quality in the sector of financial services is specific, given that the competition is intensified around the supply of qualitative services to the customers. The need to become competitive towards quality and to ameliorate customer service becomes more urgent (Rees, 1995). Service quality and customer satisfaction have been recognized as a strategic means for the successful confrontation of the competition among the financial institutions. Any change in the market could affect the preferences and decisions of the customers (Michelis et al., 2001). It is supported that the banks should be oriented strictly towards their customers in order to be more competitive and profitable and to attract more customers. The cost that is demanded for the acquisition and maintenance of the customers by an institution that provides bad quality services are much larger than the cost of another one that offers high quality services (Sohal, 1994).

√ The bank should be oriented to new data and revise its strategy, in order to succeed in the localization of sectors with development possibilities, high level of service, appreciation and confidence from the customers' point of view. Another strategy proposal, during the policy that was followed in the present study is the negotiation towards the loans' interest rates or even the rendering of best insurance in the market of a loan to the regular and stable customers of the bank, concerning the loan supply. Moreover, the reduction of the loans' interest rates, as well as the reduction of the demanded information about the economic situation of the candidate customer, the accuracy and conducting speed of the transactions contribute to the attraction of a larger number of customers.

√ The operation of the European banking institutions in an environment of low interest rates and intense competition has contributed to the reduction of the operating margins and enforced the banks to focalize new activities based on the provision of income such as insurance, the allowance for consulting and investment services, portfolio management etc. In this effort several commercial banks have acquired either insurance companies or smaller investment banks aiming at the enrichment of their offered services and simultaneously at the diversification of their income sources. Taking into account that the financial institutions in Europe will continue to operate in a more competitive environment, the mergers and acquisitions will continue aiming at the amelioration of their profitability and the control of larger market shares. Thus, banking institutions will take advantage of the scale economies, the possibilities of co operations and reduction of the cost of the offered services and they will create strategies that confront efficiently the need for new products, contribute to the optimal customer service, as well as to the most efficient geographical expansion to cover new markets.

√ Finally, a significant issue in the optimal management of assets and liabilities is the management of the operating cost and technology. The complexity of banking activities has gradually increased risks that are connected to the wide difficulty of surveillance, regulation and control of the set of activities in banking institutions, that is the operating risk. The management of the operational risk constitutes a discrete sector of operations activity with distinct administrative structure, tools and procedures. The restructurization of the operations structure and the input of new technological structures contribute to a low cost ban. This implies the acquisition of an important competitive advantage against the other banks, which is expressed through the policy of attracting loans and deposits through lower loans' interest rates and higher deposits' interest rates.

Chapter 5

Conclusions and future perspectives

1. SUMMARY OF MAIN FINDINGS

Asset Liability Management (ALM) is an important issue in the financial risk management, whose significance has been pointed out by various researchers through the development of scientific tools and methodologies for their evaluation.

More precisely, ALM is a great dimension of risk management, in which the exposition to various risks is minimized maintaining the appropriate combination of assets and liabilities in order to satisfy the goals of the firm. Its main purpose is the maximization of the firm profits and the minimization of the risk.

Applications of the above model are encountered insurance companies, banks, portfolios, pension funds and generally in financial issues.

The present study has dealt with the study of asset liability management in commercial banks. The reasons that have led us to the study of this sector of ALM are the following:

√ The intensifying international competition and the integration of the global market, the all embracing technological changes which allow banks to decentralize the decision taking increasing the speed of their response to the changes of the market, the social and demographical changes that led to the increase of the mobility and the flexibility of the labor market increased the uncertainty and the risk and modulated conditions of intense competition, which cause important changes in the Greek banking environment.

√ The liberation of the financial markets, the internationalization of the economic activity and the adoption of Euro accelerated the generation activities of the development of a real European market of financial services and pointed out the necessity of enlargement of the European financial institutions in order to procure the possibility of covering sufficiently the new widening market and consequently to be protected from eventual aggressive takeovers.

During these opportunities and threats in the international environment and in order to be competitive, banks select strategies that are centered on the substantial amelioration of their productivity, as well as on the quality of their assets and liabilities management. The main problem that arises is: which should be the composition of asset and liability of a bank given the returns and costs in order to succeed larger efficiency in the bank profits. This question led to studies regarding the optimal management of bank assets and liabilities, risk, return and liquidity.

Many are the applications and the models that were developed around the ALM. There are deterministic and stochastic models. The deterministic models use linear programming and are computationally feasible for large problems, while the stochastic models refer to the programming under uncertainty and present computational difficulties. There is an extensive bibliographical review of ALM techniques in Chapter 2. Each model presents discreteness and most of them are applied in the sector of insurance companies, portfolio management or mutual funds management and pension funds. The relative models that concern the banking section and have been developed are fewer.

Based on the above the present study presents the concepts of ALM and its applications, as well as an extensive reference to the economic role of the financial institutions and the management policy of the commercial banks. Since the development of an ALM model takes place in an environment of changing interest rates, the reference to the risk management, the risks of the financial institutions and their measurement techniques becomes essential. Moreover, the multicriteria consideration of the ALM problem is taken into account. The objective of the present book is the development of a multicriteria programming model for the optimal management of bank asset and liability management in an environment of changing interest rates.

More specifically, taking into consideration the data of the financial statements of a large commercial bank of Greece for the economic year *t* a goal programming model is developed. This model aims at the determination of the asset and liability data of the year *t+1* based on the preferences and demands of the bank managers and resolves the problem presenting a series of future forecasts and determinations of the structure of the balance sheet accounts based on environmental, solvency, liquidity, demands, loans constraints as well as constraints of asset liability return.

While the previous studies in the banking field centralize their interest on the goals of securities and financial derivatives, the proposed methodology refers to the categories of asset and liability accounts as they are presented to the departmental accounting plan of banks. Moreover, as it is mentioned in Chapter 1, since the ALM of a bank is connected swiftly to the interest rate risk, it is obvious that the changes in the values of interest rates contribute to the variability of the structure of the assets and liabilities. The present research takes into account the constraint goal of the assets and liabilities return, which constitutes a basic unknown element to the developed goal programming problem, since the accounts of assets and liabilities are related to the parameters regarding the bonds, loans and deposits interest rates. An interest rate simulation analysis takes place in order to encounter successfully the uncertainty regarding the variables of interest rates. The proposed methodology, apart from the determination of the future asset liability management strategy for the year *t+1* based on the bank's financial statements for the economic year *t*, proceeds to a sensitivity analysis as far as the levels of goal priorities are concerned. Various priority levels for the satisfaction of goals and constraints depending on the demands of bank managers are assumed. Taking into account the consideration of scenarios for the values of interest rates, different optimal solutions for the variables of asset liability data are obtained. Finally, computing the optimal values of assets and liabilities that were obtained for the year *t+1* from the development of the goal programming model, the forecast and determination of assets and liabilities for the year *t+2* takes place.

The results of the research and their comparison with those of the actual strategy that the bank has followed for the year *t*, indicate that the values of the variables concerning the accounts of loans and demands differ significantly from the values of the actual solution (Chapter 4). This is due to the fact that specific constraints have been imposed on the loans and deposits goals during the research. Moreover, since the major source of funds for a commercial bank is the loans and deposits accounts, the balance sheet structure and its optimal management is the result of the structure of the loans and deposits accounts.

The analysis of results based on the bank policy, which was applied at the present volume for the modulation of the constraints and goals of the goal programming model leads to the decision taking for the optimal ALM strategy. The strategic choices of the bank managers impose the attracting of a larger number of customers, the increase of the variety of their activities, as well as their capability of providing high quality services.

Finally, given the fact that the strong competitive urges of the last decade result in the creation of an unstable and changing environment in the European and in general in the international bank market, the banks encounter the great changes that originate from the release of the banking system from the normative adjustments, the opportunities and threats of the extensive mergers and acquisitions that take place as well as the internationalization of the activities and markets. The banks, under the pressure of these phenomena in combination with the increasing uncertainty and the risk that emerge due to the above, are enforced to become more competitive and to develop long term strategies in order to encounter the needs for new products, to enlarge their geographical presence and their participation in new markets. Thus, they are enforced to proceed to the most efficient management of their financial risks and the optimal structure of their assets and liabilities. This is accelerated through the goal programming model, which provides the decision maker / bank manager with the possibility of determining several information data in relation to their preferences to the pre-specified goals of the problems. This information could concern the determination of goals and constraints that should be ameliorated, the evaluation of several solutions that are derived based on the info determined during the solution procedure (Doumpos, and Zopounidis, 2001). With the determination of information a new optimal solution is detected. This solution constitutes the basis for the continuation of the same procedure until the optimal solution is detected, which corresponds to the preferences and the policy that the administrative council of the bank follows or wishes to follow.

2. ISSUES FOR FURTHER RESEARCH

Although the development of the specific ALM model gives the possibility to banks to proceed to various scenarios of their future economic process, aiming at the management of risks, which emerge from the changes of the market parameters, research in this field presents great prospects for further research.

The most important prospects and the corresponding future research directions are detected to the followings:

- *Investigation of the determination of the exogenous factors and the economic parameters of the market.* In the present volume, the development of the bank ALM model into an environment of changing interest rates takes place. The most significant factors that affect the changing of the composition of assets and liabilities are concerned with loans, deposits and bonds interest rates, since they refer to the corresponding categories of accounts that contribute to the basic source of funds for a commercial bank. Besides the interest rate risk, the determination of the inflation factor has an important role in the structure of the bank balance sheet and thus its effect on the balance sheet accounts is worth further research and studies.
- *Long term investigation taking into consideration the categories of accounts.* The proposed methodology of the present volume refers to the categories of the asset and liability accounts as they are presented to the departmental accounting plan of banks. Parallel to the analysis of these accounts, the investigation and further analysis of the outline accounts presents increased interest. This means that the outline accounts of bonds should be taken into account. But these data constitute a confidential and secret source of the bank and are difficult to be consolidated. It is worth noting that the consolidation difficulty of the above data has not rendered possible this sort of study in the present volume. Taking into account the different maturity dates of the outline accounts of bonds, as well as the changes to the values of interest rates it is possible to investigate the Monte Carlo simulation model which is applied in the present volume through the consideration of scenarios for the determination of the interest rate risk. This further research provides the possibility for a more precise and explicit determination of the bank balance sheet structure in an environment of changing interest rates and different maturities.
- *Investigation of the possibility of determination of the bank financial risks through the application of financial derivatives.* The financial derivatives allow the financial managers to minimize the borrowing cost of the financial institutions diachronically for given risk profiles. Several studies prove that the use of financial derivatives is an endogenous factor of the strategic management of risk for a firm or for a financial institution. Guay (1999) in his research observed a considerable reduction of the risk, when the firm uses the financial derivatives. The investigation of the specific subject has particular interest for the study of the strategic reduction of the risk with the development of models and

strategic reduction of the risk with the development of models and the application of financial derivatives.

- *Investigation of the application of the bank ALM system to the European and international environment.* The creation of the uniform monetary system brought changes to the economic environment and contributed to the extension of the net and size of banks, resulting in the amelioration of the operating efficiency of the banks in the European and international environment, in which they are active. Quite interesting is the study of application of the proposed model to the data of a large commercial bank of the European Union.

The investigation of the above future research directions will contribute substantially to a more complete definition of the ALM of commercial banks, through an integrated information system which gives the possibility to the decision maker to proceed to various scenarios of the economic process of the bank in order to monitor its financial situation and to determine the optimal strategic implementation of the composition of assets and liabilities.

References

Alexander, C. (1999), *Risk Management and Analysis-Volume1: Measuring and modeling financial risk*, John Wiley and Sons, England.

Alvord, Ch. H. III (1983), "The Pros and Cons of goal programming: a reply", *Computers and Operations Research*, 10/1, 61-62.

Aouni, B. and Kettani, O. (2001), Goal programming model: A glorious history and a promising future, *European Journal of Operational Research*, 133/2, 225-231.

Arthur, J.L. and Ravindran, A. (1978), "An Efficient Goal Programming Algorithm using constraint partitioning and variable elimination", *Management Science*, 24/8, 867-868

Arthur, J.L. and Ravindran, A, (1980), "A Branch and Bound Algorithm with Constraint Partitioning for Integer Goal Programming Problems", *European Journal of Operational Research*, 4/6, 421-425.

Ballestero, E. and Romero, C. (1998), *Multiple Criteria Decision Making and its Applications to Economic Problems*, Edition Kluwer Academic Publishers.

Baston, R.G. (1989), "Financial planning using goal programming", *Long Range Planning*, 22/17, 112-120.

Beale, E, (1967), "Numerical Methods" in Abadie, J. (Ed) *Nonlinear Programming*, North Holland, Amsterdam.

Beatty, A. (1999), "Assessing the use of derivatives as part of a risk-management strategy", *Journal of Accounting and Economics*, 26, 353-357.

Bell, D. (1995), "Risk, return, and utility", *Management Science*, 40, 23-30.

Benayoun, R., De Montgolfier, J., Tergny, J. and Larichev, O. (1971), "Linear programming with multiple objective function: Stem method (STEM)", *Mathematical Programming*, 1/3 , 366-375.

Bessler, W. and Booth G.G. (1994), "An interest rate risk management model for commercial banks", *European Journal of Operational Research*, 74, 243-256.

Birge, J.R. and Louveaux, F.V. (1988), "A multicut algorithm for two stage stochastic linear programs", *European Journal of Operations Research*, 34, 384-392.

Bitran, G.R. and Novaes, A.G, (1973), "Linear Programming with a Fractional Objective Function", *Operations Research*, 21, 22-29.

Booth, G.G. (1972), "Programming Bank Portfolios under Uncertainty: An Extension", *Journal of Bank Research* 2, 28-40.

Box, J., (1965), "A New Method of Constrained Optimization and a Comparison with Other Methods", *Computer Journal*, 8, 42-52.

Bradley, S.P., and Crane, D.B. (1972), "A Dynamic Model for Bond Portfolio Management", *Management Science* 19, 139-151.

Breiman, L. (1961), "Optimal gambling system for favorable games", *Proceedings of the 4th Berkeley Symposium on Mathematical Statistics and Probability* 1, 63-68.

Brennan, M.J. and Schwartz, E.S. (1982), "An equilibrium model of bond pricing and a test of market efficiency", *Journal of Financial and Quantitative Analysis* 17, 75-100.

Brennan, M.J., E.S. Schwartz, and R. Lagnado (1997), "Strategic Asset Allocation". *Journal of Economic Dynamics and Control*, 21/8-9, 1377-1403.

Brennan, M.J. and Schwartz, E.S. (1998), *The use of Treasury bill futures in strategic asset allocation programs*, Worldwide Asset and Liability Modelling, Ziemba, W. and Mulvey, J. (Eds.), Cambridge University Press, 205-228.

Brodt, A.I. (1978), "Dynamic Balance Sheet Management Model for a Canadian Chartered Bank", *Journal of Banking and Finance*, 2/3, 221-241.

Bryson, N.A. and Gass, S.I. (1994), "Solving Discrete Stochastic Linear Programs with Simple Recourse by the Duaplex Algorithm", *Computers and Operations Research*, 21, 11-17.

Carino, D.R., Kent, T., Muyers, D.H., Stacy, C., Sylvanus, M., Turner, A.L., Watanabe, K. and Ziemba, W.T. (1994), "The Russell-Yasuda Kasai Model: An Asset/Liability Model for a Japanese Insurance Company Using Multistage Stochastic Programming", *Interfaces* 24, 29-49.

Carino, D.R., Myers D.H. and Ziemba, W.T., (1998), "Concepts, technical issues and uses of the Russell-Yasuda Kasai financial planning model", Report, Frank Russell Company, January, *Operations Research*.

Carino, D.R. and Ziemba, W.T., (1998), "Formulation of the Russell-Yasuda Kasai financial planning", Report, Frank Russell Company, January, *Operations Research*.

Chambers, D. and Charnes, A. (1961), "Inter-Temporal Analysis and Optimization of Bank Portfolios", *Management Science*, 7, 393-410.

Charnes, A., Cooper, W.W. and Ferguson, R. O. (1955), "Optimal Estimation of Executive Compensation by Linear Programming", *Management Science*, 1/2, 138-151

Charnes, A. and Cooper, W.W. (1961), *Management Models and Industrial Applications of Linear Programming*, Wiley, New York.

Charnes, A. and Cooper, W.W. (1977), "Goal Programming and Multiple Objective Optimization (Part 1)". *European Journal of Operational Research*, 1/1, 39-54.

Charnes, A. and Littlechild, S.C. (1968), "Intertemporal Bank Asset Choice with Stochastic Dependence", Systems Research Memorandum no.188, The Technological Institute, Nortwestern University, Evanston, Illinois.

Charnes, A. and Thore, S. (1966), "Planning for Liquidity in Financial Institution: The Chance Constrained Method", *Journal of Finance*, 21/4, 649-674.

Cheng, P. (1962), "Optimum Bond Portfolio Selection", *Management Science*, 8/4, 490-499.

Chua, J.H. and Woodward, R.S. (1983), "J.M. Keyne's investment performance: a note", *Journal of Finance*, 38, 232-235.

Cohen , K.J. and Hammer, F.S. (1967), "Linear Programming and Optimal Bank Asset Management Decisions", *Journal of Finance*, 22, 42-61.

Cohon, J.L., (1978), *Multiobjective Programming and Financial Planning*, Academic Press, New York, NY.

Cohen, K.J. and Thore, S. (1970), "Programming Bank Portfolios under Uncertainty", *Journal of Bank Research*, 2, 28-40.

Commission of the European Communities, "Proposal for directive concerning the solvency ratio for credit institutions", COM (89) 239 Final, 1989.

Cosset, J.C., Siskos, Y. and Zopounidis, C. (1992), "Evaluating country risk: A decision support approach", *Global Finance Journal* , 3/1, 79-95.

Crane, B. (1971), "A Stochastic Programming Model for Commercial Bank Bond Portfolio Management", *Journal of Financial Quantitative Analysis* 6, 955-976.

Crane, B., Knoop, F. and Pettigrew, W. (1977), "An Application of Management Science to Bank Borrowing Strategies", *Interfaces*, 8/1, Part 2, 70-81.

Dantzig, B. and Glynn, P. (1990), "Parallel Processors for Planning under Uncertainty", *Annals of Operations Research 22*, 1-21.

Dantzig, B. and Infanger, G. (1993), "Multi-stage stochastic linear programs for portfolio optimization", *Annals of Operations Research*, 45, 59-76.

Dantzig, G.B. and Wolfe, P. (1960), "Decomposition principle for linear programs", *Operations Research*, 8/1, 101-111.

Das, S. (1998), *Risk Management and Financial Derivatives: A Guide to the Mathematics*, Irwin Library of Investment and Finance, McGraw-Hill, New York.

Dauer, J.P. and Krueger, R.J. (1977), "An Iterative Approach to Goal Programming", *Operational Research Quarterly*, 28/ 3, 671-681.

Derwa, L. (1972), "Computer Models: Aids to management at Societe Generale de Banque", *Journal of Bank Research*, 4/3, 212-224.

Doumpos, M. and Zopounidis C. (2002), *Multicriteria Decision aid classification methods*, Kluwer Academic Publishers

Dyer, J.S., (1972), "Interactive Goal Programming", *Management Science*,19/1, 62-70.

Eatman, L. and Sealey Jr. (1979), "A Multi-objective Linear Programming Model for Commercial bank Balance Sheet Management", *Journal of Bank Research*, 9, 227-236.

El-Dash, A.A. and Mohamed, M.B., (1992), "Sequential Duality Method for Solving Polynomial Goal Programming Problems", *Egyptian Computer Journal*, 20/1, 12-38.

Eppen, G.D. and Fama, E.F. (1971), "Three Asset Cash Balance and Dynamic Portfolio Problems", *Management Science*. 17, 311-319.

Fang, S. and Puthenpura, S. (1993), *Linear Optimization and Extensions: Theory and Algorithms*, Prentice Hall, Englewood Cliffs, NJ.

Fielitz, D. and Loeffler, A. (1979), "A Linear Programming Model for Commercial Bank Liquidity Management", *Financial Management*, 8/3, 44-50.

Finnetry, D. (1988), "Financial Engineering in corporate finance: An overview", *Financial Management*, 17, 14-33 .

Fortson, C. and Dince, R. (1977), "An Application of Goal Progamming to Management of a Country Bank", *Journal of Bank Research*, 7, 311-319.

Gass, Saul I., (1987) "The Setting of Weights in linear goal-programming problems", *Computers and Operations Research*, 14/ 3, 227-230.

Giokas, D. and Vassiloglou, M. (1991), "A Goal Programming Model for Bank Assets and Liabilities", *European Journal of Operations Research*, 50, 48-60.

Glynn, P.W. and Iglehart, D.L. (1989), "Importance sampling for stochastic simulations", *Management Science*, 35, 1367-1391.

Goicoechea, A., Hansen, D.R. and Duckstein, L. (1982), *Multiobjective decision analysis with engineering and business applications*, John Wiley and Sons, New York.

Green, P.E. and Srinivasan, V., (1990) "Conjoint Analysis in Marketing: New Developments with Implications for Research and Practice", *Journal of Marketing*, 54,/4, 3-19.

Griffith, R. and Stewart, R., (1961), "A Nonlinear programming technique for the optimization of continuous processing systems", *Management Science*, 7, 370-392.

Griffiths, W. E., Carter Hill, R. and Judge, G.G. (1992), *Learning and Practicing Econometrics*, John Wiley and Sons, New York.

Grubmann, N. (1987), "BESMOD: A Strategic Balance Sheet Simulation Model", *European Journal of Operations Research* 30, 30-34.

Guay, R. W. (1999), "The Impact of derivatives on firm risk: An empirical examination of new derivative users", *Journal of Accounting and Economics*, 26, 319-351.

Güven, S. and Persentili, E. (1997), "A Linear Programming Model for Bank Balance Sheet Management", *Omega, International Journal of Management Science*, 25/4, 449-459.

Hakansson, N.H. (1972), "On optimal myopic portfolio policies with and without serial correlation", *Journal of Business*, 44, 324-334.

Hakansson, H. and Ziemba, W. (1995), *Capital growth theory*, In Finance, R.A. Jarrow, V. Maksimovic and W. T. Ziemba, (Eds.), 123-144.

Hannan, Edward L. (1980) "Nondominance in Goal Programming", *INFOR*, 18/4, 300-309.

Hannan, Edward L. (1985), "An Assessment of some criticisms of goal programming", *Computers and Operations Research*, 12/ 6, 525-541.

Harrald, J., Leotta, J., Wallace, W.A. and Wendell, R.E. (1978) "A Note on the Limitations of Goal Programming as Observed in Resource Allocation for Marine Environmental Protection", *Naval Research Logistics Quarterly*, 25/4, 733-739

Hicks, J.R. (1939), *Value and Capital*, Cambridge: Oxford University Press.

Hogan, M. (1993), "Problems in certain two-factor term structure models", *Annals of Applied Probability*, 3, 576.

Houthakker, H.S., (1968), *Normal Backwardation*, In Value, Capital and Growth: Papers in Honor of sir John Hicks, J.N. Wolfe (Ed.), Chicago, Aldine Publishing Co, 193-214.

Hwang, C.L., Masud, A.S.M., Paidy, S.R. and Yoon K.I. (1979), *Multiple objective decision making-Methods and applications: A state-of-the-Art Survey*, Springer-Verlag, New York.

Ignizio, J.P. (1976), "An Approach to the Capital Budgeting Problem with Multiple Objectives", *The Engineering Economist*, 21/4, 259-272.

Ignizio, J.P. (1982), *Linear Programming in Single and Multiple Objective Systems*, Prentice-Hall, Englewood Cliffs, NJ.

Ignizio, James P. (1985) "An Algorithm for Solving the linear goal programming problem by solving its dual", *Journal of the Operational Research Society*, 36/ 6, 507-515.

Ijiri, Y. (1965), *Management goals and Accounting for Control*, Rand-McNally, Chicago, IL

J.P. Morgan and Company (1997), *Credit Metrics-technical document-The benchmark for understanding credit risk*, New York: J.P. Morgan and Company.

Jackson P., Maude D. and Perrudin W. (1997), "Bank Capital and Value at Risk", *Journal of Derivatives*, 4/3, 73-89.

Jorion, P. (1997), *Value at Risk: The New Benchmark for Controlling Derivatives Risk*, Mc Graw Hill.

Kallberg, J.G., White, R.W., and Ziemba, W.T. (1982), "Short Term Financial Planning under Uncertainty", *Management Science*, 28, 670-682.

Kallberg, J.G., and Ziemba, W.T. (1983), "Comparison of alternative utility functions in portfolio selection problems", *Management Science*, 29, 1257-1276.

Keeney, R.L. and Raiffa, H. (1993), *Decisions with Multiple Objectives: Preferences and Value Trade-offs*, Cambridge University Press, Cambridge.

Kelly, J. (1956), "A new interpretation of information rate", *Bell system Technology Journal*, 35, 917-926.

Kira, D.S. and Kusy, M.I. (1990), "A Stochastic Capital Rationing Model", *Journal of Operational Research Society*, 41, 853-863.

Kischka, P., (1984) "Bestimmung optimaler Portfolios bei Ungewibheit", *Mathematical Systems in Economics* 97, Athenaum/Hanstein, Konigstein.

Kochan, T.A. and Dyer, L. (1993) "Managing transformational change: the role of human resource professionals", *The International Journal of Human Resource Management*, 4/3, 569-590.

Komar, R. (1971), "Developing a Liquidity Management Model", *Journal of Bank Research*, 38-53.

Kosmidou, K. and Spathis, Ch., (2000), "Euro and profitability of Greek Banks", *European Research Studies Journal*, 3/3-4, 43-56.

Kosmidou, K. and Zopounidis, C. (2001), *Bank Asset Liability Management Techniques: An Overview*, Zopounidis C., P.M. Pardalos, G. Baourakis (Eds.), Fuzzy Set Systems in Management and Economy, World Scientific Publishers, 255-268.

Kosmidou K. and Zopounidis C. (2002), *A Multiobjective Methodology for Bank Asset Liability Management* , Pardalos P., V. Tsitsiringos (Eds.), Financial Engineering, e-Commerce and Supply Chain, Kluwer Academic Publishers, 139-150.

Korhonen, P. (1988), "A visual reference direction approach to solving discrete multiple criteria problems", *European Journal of Operational Research*, 34,152-159.

Korhonen, P. and Wallenius, J. (1988), "A Pareto race", *Naval Research Logistics*, 35, 615-623.

Kupiec, P. (1995), "Technique for Verifying the Accuracy of Risk Management Models", *Journal of Derivatives*, 3/2, 73-84.

Kusy, I. M., Ziemba, T. W., (1986), "A Bank Asset and Liability Management model", *Operations Research*, 34/3, May-June 1986, 356-376.

Kvanli, A.H. (1980), "Financial planning using goal programming-OMEGA", *The International Journal of Management Science*, 8/2, 207-218.

Langen, D., (1989), *Strategic Bank Asset Liability Management,* European University Studies.

Lee, Cheng,F. (1985), *Financial Analysis and Planning: Theory and Application*, Addison-Wesley Publishing Company.

Lee, S.M. (1972), *Goal Programming for Decision Analysis*, Auerbach Publishers, Philadelphia, PA

Lee, S.M. (1983), "A Revised Iterative Algorithm for Decomposition Goal Programming", *International Journal of Systems Science*, 14/12, 1383-1393.

Lee, S.M. (1985a), "A Gradient Algorithm for Chance Constrained Nonlinear Goal Programming", *European Journal of Operational Research*, 22/3, 359-369.

Lee, S.M. and Chesser, D.L. (1980), "Goal programming for portfolio selection", *The Journal of Portfolio Management*, 6, 22-26.

Lee, S.M. and Rho, B.H. (1979a) "The Binary Search Decomposition in a Decentralized Organization", *Theory and Decision*, 11/4, 353-362.

Lee, S.M and Rho, B.H. (1979b), "The Modifieed Kornia-Liptak Decomposition Algorithm", *Computers and Operations Research*, 6/1, 39-45.

Lee, S.M. and Rho, B.H. (1985), "A Multicritieria Decomposition Model for Two-Level, Decentralized Organization", *International Journal of Policy and Information*, 9/1, 119-134.

Lee, S.M. and Rho, B.H. (1986), "Computational Experience with the Dantzig-Wolfe and Kornai-Liptak Decomposition Algorithm", *International Journal of Policy and Information*, 10/1, 83-94.

Lee, S.M. and Lerro, A.J. (1973), "Optimizing the portfolio selection for mutual funds", *Journal of Finance*, 28, 1086-1101.

Lifson, K.A. and Blackman, B.R. (1973), "Simulation and Optimization Models for Asset Deployment and Funds Sources Balancing Profit Liquidity and Growth", *Journal of Bank Research*, 4/3, 239-255.

Lofti, V., Stewart, T.J. and Zionts, S. (1992), "An aspiration-level interactive model for multiple criteria decision making", *Computers and Operations Research*, 19, 677-681.

Macaulay, F.R. (1938), *Some Theoretical Problems Suggested by the Movement of Interest Rates, Bond Yields and Stock Prices Since 1856*, New York: National Bureau of Research.

Markowitz, H.M. (1952), "Portfolio Selection", *Journal of Finance*, 7/1, 77-91.

Markowitz, H.M. (1959), *Portfolio Selection, Efficient Diversification of Investments*, John Wiley and Sons, New York.

Marto, B., (1964), "Hyperbolic Programming", *Naval Research Logistics Quarterly*, 11, 135-155.

Merton, R.C. (1969), "Lifetime portfolio selection under certainty: the continuous time case", *Review of Economics and Statistics* 3, 373-413.

Merton, R.C. (1990), *Continuous-Time Finance*. Blackwell Publishers.

Merton, R.C. (1998), *Optimal Investment Strategies for University Endowment Funds*, Worldwide asset and liability modeling, Cambridge University Press.

Michelis, G., Grigoroudis, E., Siskos, Y., Politis, Y. and Malandrakis Y., (2001), Customer satisfaction measurement in the private bank sector", *European Journal of Operational Research*, 130/2, 347-360.

Michnik, J., (2002), *Multiobjective Analysis of a Financial Plan in a Bank*, Advances in Soft Computing-Multiple Objective and Goal Programming, Trzaskalik T. and Michnik, J. (Eds.), Physica-Verlag, New York, 351-361.

Miller, C. (1963), "The Simplex Method for local separable programming" in Graves, R.L. and Wolfe, P. (Eds), *Recent Advances in Mathematical Programming*, McGraw-Hill, New York, NY.

Min, H. and Storbeck J., (1991), "On the Origin and Persistence of Misconception in goal programming" *Journal of the Operational Research Society*, 42/1, 301-312.

Moore, L.J., Taylor, B.W.III, Clayton, E.R. and Lee, S.M., (1978) "Analysis of a Multi-Criteria Project Crashing Model", AIIE Transactions, vol. 10, no2, 163-169

Mulvey, J.M. (1996), "Generating scenarios for the Towers Perrin investment system", *Interfaces* 26, 1-15.

Mulvey, J.M. and Chen Z., (1996), *An empirical evaluation of the fixed-mix investment strategy*, Princeton University Report SOR-96-21.

Mulvey, J.M., Correnti, S., and Lummis, J. (1997), *Total integrated risk management: insurance elements*, Princeton University Report, SOR-97-2.

Mulvey, J.M., Rosenbaum, D.P. and Shetty, B. (1997), "Strategic financial risk management and operations research", *European Journal of Operational Research*, 97, 1-16.

Mulvey, J.M. and Crowder, H.P. (1979), "Cluster Analysis: An Application of Lagrangian Relaxation", *Management Science*, 25/4, 329-340 .

Mulvey, J.M. and Thorlacius, A.E. (1998), *The Towers Perrin global capital market scenario generation system*, Worldwide Asset and Liability Modelling, Ziemba, W. and Mulvey, J. (Eds.), Cambridge University Press, 286-312.

Mulvey, J.M. and Vladimirou, H. (1989), "Stochastic Network Oprimization Models of Investment Planning", *Annals of Operations Research* 20, 187-217.

Mulvey, J.M. and Vladimirou, H. (1992), "Stochastic Network Programming for Financial Planning Problems", *Management Science* 38, 1642-1663.

Nanda, J., Kothari, D.P. and Lingamurthy, K.S. (1988), "Economic Emission Load Dispatch through Goal Programming Techniques", *IEEE Transactions on Energy Conversion*, 3/1, 26-39

Newton, K. (1985), "Interpreting Goal Attainment in chance-constrained goal programming", *OMEGA*, 13/1, 75-78.

Oguzsoy C.B. and Güven S. (1997), "Bank Asset and Liability Management under uncertainty", *European Journal of Operational Research* 102, 575-600.

Olson, D.L. and Swenseth, S.R., (1987), "A Linear Approximation for chance-constrained programming", *Journal of the Operational Research Society*, 38/ 3, 261-267.

Pallaschke, D. and Vollmer, K. (1985), "A Partial Approach for the Optimization of Balance Sheet Structures under Special Risk-Functions", *Proceedings of the 3rd Symposium on Money, Banking and Insurance*, 699-710.

Papoulias, G.P., (1982), *Financial management and policy*, Editions Th. Tirovola, Athens (in Greek).

Pareto, V. (1896), *Cours d'Economie Politique*, Lausanne.

Perold, A.F. and Sharpe, W.F., (1988). "Dynamic strategies for asset allocation". *Financial Analysts Journal*, January, 16-27.

Pogue, G.A., and Bussard, R.N. (1972), "A Linear Programming Model for Short Term Financial Planning under Uncertainty", *Sloan Management Review* 13, 69-98.

Porter, R., (1962), "A Model of Bank Portfolio Selection", *Yale Economic Essays*, 2/1, 323-529.

Pyle, D.H. (1971), "On the Theory of Financial Intermediation", *Journal of Finance* 26, 737-746.

Ress, C. (1995), "Quality management and HRM in the service industry: some case study evidence", *Employee Relations*, 17/3, 99-109.

Robertson, M. (1972), *A Bank Asset Management Model*, Eilon, S. and Fowkes, T.R. (Eds.), Applications of Management Science in Banking and Finance, Gower Press, Epping, Essex, 149-158.

Robichek, A.A., Techroew, D. and Jones, J.M., (1965), "Optimal Short Term Financing Decision", *Management Science*, 12/1, 1-36.

Robinson, R.S. (1973), "BANKMOD: An Interactive Simulation Aid for Bank Financial Planning", *Journal of Bank Research*, 4/3, 212-224.

Romero, C. (1986), "A Survey of generalized goal programming (1970-1982)", *European Journal of Operational Research*, 25, 183-191.

Romero, C. (1991), *Handbook of Critical Issues in Goal Programming*, Pergamon Press, Oxford.

Rosenthal, R.E., (1983) "Principles of Multiobjective Optimization" *Decision Sciences*, 16/2, 133-152.

Ross, M. S., (1999), *An Introduction to Mathematical Finance: Options and Other Topics*, Cambridge University Press, United Kingdom.

Rotando, M. and Thorp, E. (1992), "The Kelly criterion and the stock market", *American Mathematical Monthly* (December), 992-1032.

Ruefli, T. W. (1971), "A Generalized Goal Decomposition Model", *Management Science*, 17/8, B505-B518.

Saaty, T.L. (1980), *The Analytic Hierarchy Process*, McGraw-Hill, New York, NY.

Saber, H.M. and Ravindran, A. (1993), "Nonlinear goal programming theory and practice: a survey", *Computers and Operations Research*, 20/3, 275-292.

Samuelson, P. (1969), "Lifetime portfolio selection by dynamic stochastic programming", *Review of Economics and Statistics* (August), 239-246.

Saunders, A. (2002), *Financial Institutions Management*, McGraw Hill.

Schniederjans, M.J., (1984), *Linear Goal Programming*, Petrocelli Books, Princeton, NJ

Schniederjans, M.J. (1995), *Goal Programming: Methodology and Applications*, Kluwer Academic Publishers, Norwell, USA.

Seshadri, S., A. Khanna, F. Harche, and Wyle R. (1999), "A method for strategic asset-liability management with an application to the federal home loan bank of New York", *Operations Research*, 47/3, 345-360.

Sharma, J.K., Sharma, D.K. and Adeyeye, J.O. (1995), "Optimal portfolio selection: A goal programming approach", *Indian Journal of Finance and Research*, 7/2,67-76.

Siskos, J. and Despotis, D.K. (1989), "A DSS oriented method for multiobjective linear programming problems", *Decision Support Systems*, 5, 47-55.

Smithson, W. Ch. (1998), *Managing Financial Risk: A guide to derivative products, financial engineering and value maximization*, McGraw-Hill, New York.

Sohal, A.S. (1994), "Managing service quality: developing a vision and a strategy", *Total Quality Management*, 5/6, 367-374.

Slowinski, R. and Teghem, J. (Eds), (1990), *Stochastic versus Fuzzy Approaches to Multiobjective Mathematical Programming under Uncertainty*, Kluwer Academic Publishers, Dordrecht.

Spathis, Ch., Kosmidou K. and Doumpos, M., (2002), "Assessing Profitability Factors in the Greek Banking System", *International Transactions in Operational Research*, 9/5, 517-530.

Spronk, J. (1981), *Interactive Multiple Goal Programming Application to Financial Planning*, Martinus Nijhoff Publishing, Boston.

Steuer, R.E. (1986), *Multiple Criteria Optimization: Theory, Computation and Application*, John Wiley & Sons, New York.

Steuer, R.E. and Choo, E.U. (1983), "An interactive weighted Tchebycheff procedure for multiple objective programming", *Mathematical Programming*, 26/1, 326-344.

Sweeney, D.J., Winkofsky, E.P., Roy, P. and Baker, N.R. (1978), "Composition vs Decomposition: Two approaches to modeling organizational decision processes", *Management Science*, 24, 1491-1499.

Tayi, G. and Leonard, P. (1988), "Bank Balance Sheet Management: An Alternative Multi-Objective Model", *Journal of the Operational Research Society* 39, 401-410.

Teghem, J., Dufrane, D., Thauvoye, M. and Kunsch, P.L., (1986), "STRANGE: an interactive method for multi-objective linear programming under uncertainty", *European Journal of Operational Research*, 26/1, 65-82.

Telser, L., (1955-6), "Safety First and Hedging", *Review of Economic Studies*, XXIII, 1-6.

Tobin, J., (1958), "Liquidity Preference as Behavior Toward Risk", *Review of Economic Studies*, 25/2, 65-85.

Von Neumann, J. and Morgenstern, O., (1944), *Theory of Games and Economic Behavior*, Princeton University Press.

Wets, R.J.-B, (1966), "Programming under Uncertainty: the Complete Problem", *Z. Wahrscheinlichk Verw. Gebiete* 4, 316-339.

Wets, R.J.-B, (1983), "Solving Stochastic Programs with Simple Recourse", *Stochastics*, 10, 219-242.

Wierzbicki, A.P. (1980), "The use of reference objectives in multiobjective optimization", in: G. Fandel and T. Gal (Eds.), *Multiple Criteria Decision Making: Theory and Applications*, Lecture Notes in Economic and Mathematical Systems 177, Springer-Verlag, Berlin-Heidelberg, 468-486.

Wilmott P. (2000), *Paul Wilmott on Quantitative Finance*, Volume 1 & 2, Wiley, England.

Wolf, C.R., (1969), "A Model for Selecting Commercial Bank Government Security Portfolios", *Rev. Econ. Stat.* 1, 40-52.

Zanakis, S. and Gupta S., (1985), "A Categorized bibliographic Survey of Goal Programming", *Omega*, 13/1, 211-222.

Zeleny, M. (1982), *Multiple criteria decision making*, McGraw-Hill, Inc., 281-313.

Ziemba, W.T. (1974), *Stochastic programs with simple recourse*, in P.L. Hammer and G. Zoutendijk (Eds.), Mathematical Programming: Theory and Practice, North Holland, Amsterdam, 213-273.

Ziemba, W.T., Mulvey, J.M. (1998), *Worldwide Asset and Liability Modeling*, Cambridge University Press.

Zionts, S. and Wallenius, J. (1976), "An interactive programming method for solving the multicriteria problem", *Management Science*, 22, 652-663.

Zoutendijk, G., (1960), *Methods of Feasible Directions*, Elsevier, Amsterdam

Subject Index